ORDER NUMBER EA-NMR

AIRCRAFT BONDED STRUCTURE

INTERNATIONAL AVIATION PUBLISHERS, INC.
TRAINING MANUAL

International Standard Book Number 0-89100-065-8
For sale by: International Aviation Publishers, Inc.
P.O. Box 36   1000 College View Drive
Riverton, Wyoming 82501-0036
Tel: 1 (800) 443-9250
(307) 856-1582

International Aviation Publishers, Inc.
1000 College View Drive, Riverton, Wyoming 82501-0036

Copyright 1985 by International Aviation Publishers, Inc.
All Rights Reserved
Printed in the United States of America

INTERNATIONAL
Aviation Mechanics Journal

THE Magazine For Aviation Maintenance Professionals

For seventeen years —
- **The** most informative
- **The** most quoted
- **The** most acknowledged
- **The** most frequently reprinted
- **The only** paid-subscription aviation maintenance trade magazine

Each month —
- Colorful features of historical and human interest
- Detailed component and systems analysis
- Maintenance tips from our readers in the field
- New products and literature
- Inspection aids and advice

No Other Magazine
Can Back You Up
On The Job Like The
JOURNAL!

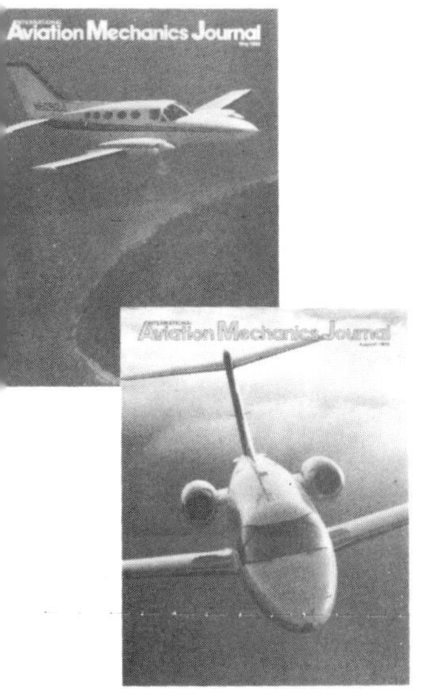

ORDER YOUR SUBSCRIPTION TODAY!

U.S. Rates:
1 year/12 issues/$16.00
2 years/24 issues/$31.00

Foreign Rates:
1 year/surface mail/$22.00
1 year/airmail/$47.00

U.S. Funds On U.S. Banks Only

CALL TOLL FREE
1-800-443-9250
In Wyoming, Or Outside Continental U.S.A.:
1 (307) 856-1582

SUBSCRIPTION ORDER FORM
PLEASE PRINT OR TYPE

S.D. # 14

COMPANY NAME________________________ PHONE___________

NAME____

ADDRESS____

CITY________________________ STATE________ ZIP_________

A&P # (if applicable)___________________________________

SIGNATURE_____________________________ DATE___________

U.S. RATES: Issued Monthly	FOREIGN RATES: (Including Mexico & Canada)
$16.00/One year	$22.00/One year/Surface mail
$31.00/Two years	$47.00/One year/Airmail
Allow 6-8 weeks for delivery	**U.S. FUNDS ON U.S. BANKS ONLY**

Please enter my subscription to the *Journal* for ____________ year(s).

☐ Enclosed please find $________ to cover my order.

☐ I wish to charge my subscription on my credit card.

☐ Am. Exp. Card #________________ Exp. Date________

☐ Master Card #________________ Exp. Date________

☐ Visa Card #________________ Exp. Date________

Be sure to check your job title and aviation affiliation below:

JOB TITLE (Check one only)

Title	Coding Symbol	Title	Coding Symbol
☐ Owner	C	☐ Engineer	D
☐ Inspector	I	☐ Instructor	B
☐ Svc. Mgr.	A	☐ A&P Mechanic	E
☐ Supervisor	G	☐ Student	H
☐ Co. Officer	J	☐ Pilot	L
☐ Avionics Tech.	F	☐ Librarian	M
☐ Other	K		

(Please specify)___________

BUSINESS OR INDUSTRY (Check One Only)

Classifications	Coding Symbol	Classifications	Coding Symbol
☐ FBO/Rep. Sta.	1	☐ School	7
☐ Airline/Air Taxi	2	☐ Ag Operator	9
☐ Mfg. or Distn.	3	☐ Self-employed	10
☐ Corp. Avia.	4	☐ Avia. Consultant	11
☐ Military	5	☐ Home Builders	12
☐ Gov. Cty.-St.-Fed.	6	☐ Flying Club	13
☐ Other	8		

(Please specify)___________

EA-FAR

YOUR COMPANION REFERENCE GUIDES

All pertinent FARs for students, mechanics, and shop personnel

Aircraft maintenance, repair, and modification practices, as published by the FAA

EA-AC 43.13 1A & 2A

For current price list, and to place your order, call toll free
1-800-443-9250 (in Wyoming, or outside continental U.S.A. 1- 307-856-1582).

☐ Please send me a **FREE** copy of your latest audio/visual and book catalog.

Name________________________ Tel. No.___________

A&P License # (if applicable)___________________________

Address____

City________________________ State________ Zip_________ -________

BUSINESS REPLY MAIL
FIRST CLASS   PERMIT #8   RIVERTON, WY 82501-9990

POSTAGE WILL BE PAID BY ADDRESSEE

INTERNATIONAL
Aviation Mechanics Journal
P.O. Box 36
Riverton, Wyoming 82501-0036

SUBSCRIPTION DEPARTMENT

NO POSTAGE
NECESSARY
IF MAILED
IN THE
UNITED STATES

BUSINESS REPLY MAIL
FIRST CLASS   PERMIT #8   RIVERTON, WY 82501-9990

POSTAGE WILL BE PAID BY ADDRESSEE

INTERNATIONAL
Aviation Publishers Inc.
P.O. Box 36
Riverton, Wyoming 82501-0036

NO POSTAGE
NECESSARY
IF MAILED
IN THE
UNITED STATES

Table of Contents

Preface

This book on *Aircraft Bonded Structure* is one of a series of specialized study guides prepared for aviation maintenance personnel, to be used with a corresponding 35MM filmstrip and recorded tape cassettes.

This series is part of a programmed learning course developed and produced by the International Aviation Publishers, Inc. (IAP), one of the largest suppliers of aviation maintenance training materials in the world. This program is part of a continuing effort to improve the quality of education for aviation mechanics throughout the world.

The purpose of each IAP training series is to provide basic information on the operation and principles of the various aircraft systems and their components.

Specific information on detailed operation procedures should be obtained from the manufacturer through his appropriate maintenance manuals, and followed in detail for the best results.

This particular manual on *Aircraft Bonded Structure* includes a series of carefully prepared questions and answers to emphasize key elements of the study, and to encourage you to continually test yourself for accuracy and retention as you use this book. A multiple choice final examination is included to allow you to test your comprehension of the total material.

Some of the words will be new to you. They are defined in the Glossary found at the back of the book.

The validity of any program such as this is enhanced immeasurably by the cooperation shown IAP by recognized experts in the field, and by the willingness of the various manufacturers to share their literature and answer countless questions in the preparation of these programs.

If you have any questions or comments regarding this program, or any of the many other programs offered by IAP, simply contact the Sales Department, International Aviation Publishers, Inc., P.O. Box 36, Riverton, WY 82501-0036; or call 1-800-443-9250 (in Wyoming or outside U.S. call 1-307-856-1582).

Introduction

The earliest airplanes designed were generally of nonmetallic construction. Spruce or bamboo formed their framework and cotton or linen covered the lifting surfaces. The occupants sat in the open until their desire for comfort brought about the enclosed fuselage. The need for streamlining resulted in the smooth aerodynamic shapes we are familiar with today.

Welded steel tubing truss fuselages met the necessary strength and weight requirements, but they were not streamlined. So in order to reduce the drag, a superstructure, covered with fabric, was built around the truss, giving it the desired shape. In the 1920's, the Lockheed Company in California developed a molding process by which thin sheets of wood were formed in heated concrete molds and glued together under heat and pressure, resulting in a truly monocoque or stressed-skin structure.

This was one of the first practical applications of what we now call "bonded structure." Airplanes constructed in this way became world famous for their high-strength, low-weight, and streamlined form. One of the better known was the Lockheed Vega, the "Winnie Mae" used by Wiley Post to establish speed, distance, and altitude records.

Aluminum alloys have replaced wood veneer as the standard material for stressed skin airplanes because of their higher strength and adaptability to mass production. Sheets of this alloy are formed in hydropresses and riveted together into the aerodynamic shapes we use today. As sheet metal gained ascendancy, the wood processes were less used, although they have never been abandoned entirely.

In the latter part of the 1970's, we find that aircraft construction uses many of the innovations of the 20's through 40's, but with new and improved materials. Instead of wood strips, we use layers of glass fiber. Casein, or milk, glue has been replaced with some of the synthetic resins such as those in the epoxy family, and rigidity has been achieved through the use of honeycomb structure.

Modern aircraft such as the Bell Jet-Ranger use many sections of bonded structure. Some use honeycomb; others, laminated glass fiber; and still others, molded plastic parts.

Bonded structure is one of the more dynamic aspects of aviation design and every day brings about new developments in adhesives, sealants, and fillers. In this book you will find not only what bonded structure is, but also the different types, uses, and results of its application.

Fig. I.1   One of the earliest applications of bonded structure was the wooden monocoque fuselage used by the Lockheed Corporation in Burbank, California in 1928, and formed in these concrete molds.

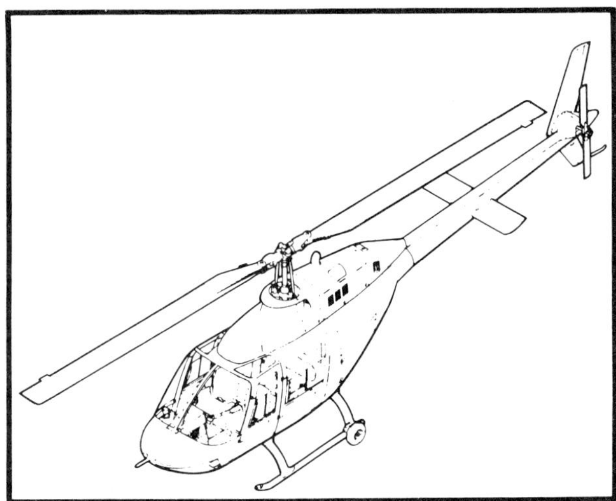

Fig. I.2   The Bell Jet-Ranger is an example of modern aircraft construction using many applications of bonded structure.

SECTION I:

What is Bonded Structure?

In its simplest form, a bonded structure exists when any two pieces of material are joined together which achieve their carry-over strength through a chemical rather than a mechanical bond. This term "chemical" comprises a wide range of terms, as we consider the various resins and adhesives used in modern aircraft construction.

Plastic has become a household word and is the center of our world of bonded structure. This type of material has been familiar to us since around 1870, when celluloid was developed. A stronger plastic was developed in 1909, when phenol and formaldehyde were mixed to start the family of phenolics we use today.

Plastics, as we use them, fall into two natural classifications, thermoplastic and thermosetting. Thermoplastic materials in their original state are hard, but become soft and pliable when heated. When soft, they can be molded or shaped, retaining this shape when they cool. Unless its heat limit is exceeded, this process can be repeated many times without damage to the material.

Thermosetting materials usually have little strength in themselves, but are used to impregnate linen, paper, or glass cloth. When formed and cured, they develop a hard consistency that will retain its shape unless damaged. Reinforced thermosetting resins are strong and lightweight and resist weathering, aging, and chemical attack.

There are four general groups of resins that make up the plastics we use in aircraft construction and repair:

1. *Natural Resins:* These include such materials as shellac, pitch, amber, asphalt, and rosin.

2. *Synthetic Resins:* These are made from petroleum, glycerol, calcium-cyanamide, benzene, urea, ethylene, phenol, and formaldehyde. Products made from synthetic resins include acrylics, nylon, vinyl, styrene, polyethylene, urea-formaldehyde, and others.

3. *Protein Plastics:* Protein plastics are manufactured from a variety of agricultural products such as peanuts, cashews, milk, coffee beans, and soybeans.

4. *Cellulose Plastics:* These are the oldest of the group and include celluloid, acetate, nitrate, ethyl-cellulose, and butyrate.

Bonded structure, as we think of it in aircraft construction, is normally of a laminated construction using thermosetting resins. One of the more popular forms of bonded structure is honeycomb, in which a core material made of thin metal foil, plastic, or fiberglass cloth in a cellular structure is covered with facings of fiberglass or sheet metal. This form is similar to that of the wax comb formed by the honeybee in which he deposits his honey.

SECTION II:

Laminated Structural Materials

Laminated may be defined as being composed of layers of materials which are firmly united. An example of this, familiar to the A&P mechanic, is the wooden propeller made of laminated birch planks. By using several layers of wood bonded together by the appropriate glue, we can construct a more uniform propeller than would be possible from a solid plank.

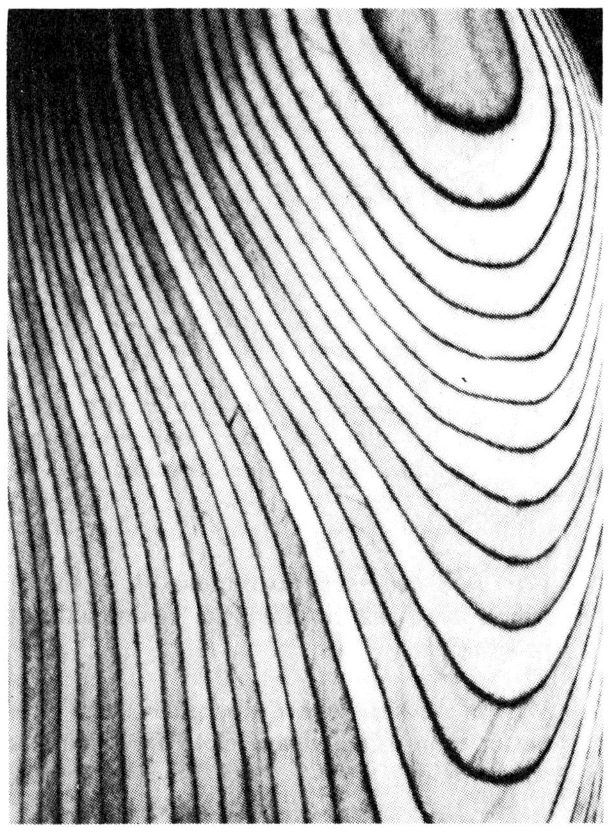

Fig. 2.1   *The laminated birch propeller is a familiar example of bonded structure.*

Using the same idea, we can form a structural material with layers of glass fiber, linen, or paper that is superior to a material that is not laminated.

Control pulleys are usually made of linen cloth impregnated with a phenolic resin. When this combination is cured under heat and pressure, it forms a superior pulley, one with enough strength that it will not be broken by the control cable, yet it will not wear the cable itself.

Fig. 2.2   *Control pulleys are made of linen cloth laminates bonded together with phenolic resin.*

Electrical insulating material which requires a great deal of mechanical strength is often made of phenolic-resin impregnated paper.

Perhaps the main reason for using bonded structure lies in the ease with which it may be manufactured and repaired, the complex shapes and forms it can assume, and its low cost in comparison with sheet metal parts. One of the combinations which has made all of these requirements attainable is glass fiber cloth and polyester resin. We will look at both of these materials separately, and then see just what they will do when they are used together.

A.   Fiberglass

First, let's consider fiberglass and the reasons for its use. Fiberglass is not a term applied to some exotic material; it is just what its name implies—fiber made of glass. From one ball of glass no larger than a marble, such as is used in schoolyard play, about 90 miles of thread, or fiber, is spun. This fiber can be woven into cloth. And herein lie the amazing properties of fiberglass cloth: Just as we have never seen a glass bottle rust, melt at less than white-hot temperatures, or corrode from some of our strongest acids, so fiberglass is just as impervious to these hazards. And even beyond this, *fiber*glass, unlike sheet glass, can be folded and bent, and like any other fiber, woven in many weights and weaves.

Fig. 2.3   Fiberglass actually consists of thin filaments, or fibers, of glass, each having the strength of glass, but because of their extremely small cross section has a flexibility not generally associated with glass.

In the manufacture of glass cloth, it is necessary to use oil. However, if cloth containing oil is used with resins, it will resist the resin and a product of low strength will result. To remove the oil, the rolls of cloth are baked in an oven at 600 degrees F. which burns off the oil (and all the other impurities), resulting in pure glass cloth. Then, in order for this cloth to have better absorption of the resins, a chrome-complex pre-wetting treatment is used. This produces a cloth which readily allows the polyester resin to flow completely around, or encapsulate, each fiber; this in turn produces a strong lamination.

Since we require the maximum strength in our cloth, it would appear that we would want a very close weave. On the other hand, the need for maximum penetration of resin requires a coarse weave, so to assure both strength and penetrability, a compromise must be struck. And a material presenting a good compromise is one having 17 to 18 threads per inch and a thickness of 0.012 to 0.014 inch, with a weight of around 9 oz. per square yard.

Fiberglass mat, somewhat similar to the cloth, is constructed of pure glass fibers which have been given the chrome-complex treatment, but, unlike the cloth, is not woven. The glass fibers are gathered and pressed loosely together, with just enough polyester resin added to hold them in place. To the eye, this fiberglass mat appears as millions upon millions of spider webs pressed together. All of these short pieces of glass fiber,

laid together at random and lightly bonded, give the mat equal strength in all directions. It resembles cloth, but it is quite a bit heavier—about 18 oz. per square yard, and thicker—roughly about 0.030 inch than its cloth counterpart.

Fiberglass mat is used for quick buildup and as a fully-impregnated core for sandwich-type structure. *Roving* is made up of glass fibers formed into a loose strand and may be used in place of mat because of its greater ease of handling.

Glass fiber tape has woven edges and is available in widths from about one eighth of an inch up.

Fig. 2.4   Fiberglass may be woven into glass cloth, packed into loose mat, or stranded into skeins of roving.

QUESTIONS

1.   Why is it necessary to remove all of the oil from fiberglass cloth in the manufacturing process?

2.   Why is a coarse-weave fiberglass cloth often better than a fine-weave cloth for laminated construction?

3.   Which is heavier, for equal strength, fiberglass mat or fiberglass cloth?

4.   What is roving?

B.   Polyester Resins

While fiberglass possesses many virtues, its greatest limitation lies in its lack of structural rigidity. Its weave is easily distorted and separated, so something must be used to give it cohesiveness and rigidity and to complement its inherent strength. Polyester resin may be used for this.

Paints and glues use oils or solvents to liquify their dry base material, and , after application, the solvent dries, leaving the base and the oils. Time and weather then dry the oil, allowing the base to chip, crack, or flake off. Polyester, on the other hand, has no oils or solvents, and it cures and hardens by a chemical reaction; not by drying.

Once polyester has cured, time and weather have little effect upon it, and when used in aircraft finishing systems, it gives a permanent glossy and practically lifetime finish. (These finishing systems are discussed in detail in the IAP text, *Aircraft Painting and Finishing*.)

Because of its unique characteristics, polyester has won its rightful place in the world of aircraft structure, and the variations in its formulation provide an almost unlimited spectrum of properties; so much, in fact, that today there are just about as many types of polyesters as there are requirements for them.

Polyester has a highly complex molecular structure, and volumes could be written on its varied molecular inner reactions. In this text, we cannot attempt an in-depth discussion of these complexities, but we will look at some of the factors to be considered when purchasing and using polyesters.

Pure polyester resin is thick and unmanageable; so it becomes necessary to add another ingredient to give it a workable texture. For this reason, a styrene monomer is added to the polyester. Styrene is not merely a thinner, which would evaporate out of the polyester; it is actually, itself, a plastic. If it is catalyzed, styrene will cure into a mass, though possessing little strength of its own. Polyester resin containing about 36-percent styrene is a good structural material. Adding more styrene only tends to weaken it, while using any less will make the polyester difficult to handle.

If the polyester resin is allowed to sit for a long period of time it will solidify, so inhibitors are added to retard this hardening process. In order, however, for the resins to cure, the inhibitors must be suppressed. This is done by using a catalyst, an agent which aids the joining together of the polyester molecules.

The material most commonly used as a catalyst is MEKP (methyl-ethyl-ketone-peroxide in dimethyl phthalate). We now have two chemicals which, when combined, will form a solid resin. The time taken for the resin to harden is too long to be practical, however, so the cure is speeded by adding an accelerator. This agent can be added to the resin either by the manufacturer, or by the technician at the time of use.

To understand the interaction of these chemicals, let's take a simplified look at what goes on during the cure. The smallest known structure which has a recognizable set of properties is the molecule. Everything we see and touch is composed of molecules and can be expressed in terms of molecular structure. Probably the most familiar structure is pure water, H_2O. This formula indicates that in any given amount of water, the atoms of hydrogen and oxygen are present in the same ratio. This same expression can be diagrammed as follows:

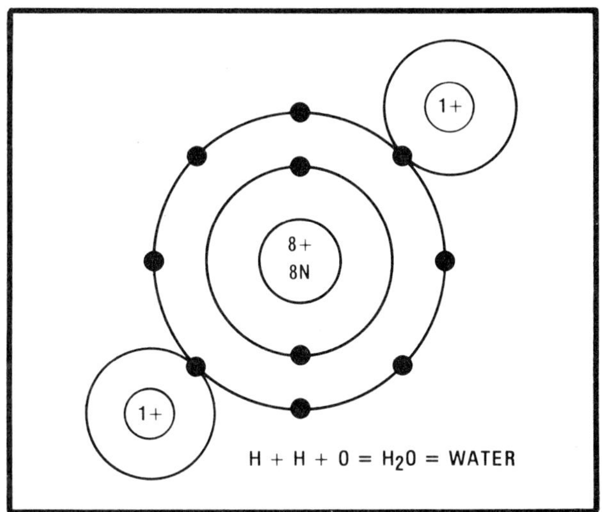

Fig. 2.5 *Water is one of the more familiar examples of a chemical compound. Because of its molecular structure, it can exist in only one form, H_2O. The more complex molecules of the resins used in bonded structure are also bound together by electron sharing.*

In most substances, the groups of molecules may be held together loosely, as in a liquid, or more rigidly, as in a solid. In the liquid state, polyester molecules are held in a weak bond, but once hardened they form solidly-linked chains which may be difficult, and sometimes impossible, to destroy.

The force which causes the loose polyester molecules to join into chains is heat. Catalyzed resins may be cured in minutes by heating the mixture, and we can see that it is advantageous if we can create internal heat for this purpose, independent of any outside source. We do this by combining the catalyst with the accelerator.

Catalysts plus accelerators result in heat. This is demonstrated by putting a few drops of accelerator on a rag and adding a few drops of catalyst. Within a short time it will be necessary to open a window to get rid of the smoke! CAUTION: The accelerator is always added to the *catalyzed resin—never* to the catalyst alone. With these chemicals producing heat for us, we have no need of an external source of heat.

QUESTIONS

5.   What is the difference between the cure process of a paint and of a polyester resin?

6.   Why is styrene added to polyester resins?

7.   What is the purpose of the catalyst added to a polyester resin?

8.   Why is an accelerator added to a polyester resin?

Resins in concentration do not transfer heat very well. If we mix two batches of resin and catalyze them equally, leave one batch in the jar and spread the other one out thin, the one in the jar will harden rapidly because of the heat trapped in the mass. This cure will take place so fast that it will cause minute fractures within the plastic. The thin sheet, on the other hand, will not cure as well because it has a large surface area exposed to the air which allows the heat to escape.

This gives us one of the first principles governing the use of polyester: The thicker the layer, the less catalyst required; the thinner the layer, the more catalyst needed. We can also prevent the loss of some of this required heat by using a resin which incorporates wax. As the resin cures, the wax migrates to the surface and forms a heat barrier.

Polyester, like all other plastics, has a shrinkage factor, and while this inherent characteristic helps in some ways, it hurts in others. For instance, when a metal structure is bonded to one made of fiberglass, the shrinkage is helpful, since the resin shrinks as it cures and provides an ever increasingly tight grip on the embedded metal. On the other hand, when hinges are installed on a large flat surface, with long strips of glass fiber and resin being used, this shrinkage will often warp the surface out of shape. One method of handling this kind of situation is to use short strips of fiberglass to bond the fasteners to the panel; or use a resin which has a smaller amount of shrinkage, a resin such as an epoxy.

For all of its good features, polyester has its drawbacks; sheets thin enough to be used for structure are very brittle, but if we add the strength of fiberglass to polyester resin, an extremely versatile material results. The fiberglass with its 300,000-psi tensile strength, tightly bonded in a polyester resin, can be formed into lightweight sheets. Any number of layers may be laminated to form a complex structure.

To insure a successful cure, certain procedures must be followed: The first and most important is to *follow the manufacturer's recommendations in extreme detail.* The second is to use a complete system provided by one manufacturer.

Mixing the polyester resin requires patience and very careful measurements. Manufacturers may specify the amount of the ingredients to be used either by weight or by volume, so be sure you understand *which* units are being used, as improper measurements will cause either a mix that will not harden at all or one that hardens too quickly.

One of the most reliable quality control indicators to use when mixing batches of polyester is the test sample. Mix enough extra resin in each batch to make an identical type of lay-up on a piece of scrap aluminum. This test sample will give a good indication of the cure-time needed, as well as its physical characteristics.

Cleaning the container used for polyesters is relatively simple: Turn the container upside down and allow the excess to drain out; then, when the resin cures, what is left will shrink away from the sides and can be easily removed.

Temperature has a major effect on polyesters; they should not be used when the temperature is below 65 degrees F. or above 85 degrees. Always store the basic materials in a cool, dry place, and use good housekeeping procedures when working with polyesters. Extreme care should be taken when the catalyst is handled, as it is a strong irritant and may cause skin reaction. The peroxide accelerator may also cause severe eye irritation or blindness.

QUESTIONS

9.   Which will cure the more rapidly, a thin layer of polyester resin, or a very heavy layer?

10.   Does polyester resin shrink or expand as it cures?

11.   Give two rules which, if followed, will ensure a successful cure of a polyester resin.

12.   What is the recommended temperature range for using polyester resins?

C.   *Epoxy Resins*

There are other resins which may take the place of polyester in laminated structure. One of these is epoxy, readily available in many forms and colors, some as thin as liquid, others thick as paste. The one most frequently encountered comes in two tubes and may be purchased at the local hardware store.

The specific properties of epoxy which make it useful for bonded structure are its low percentage of shrinkage, its high strength for its weight, its exceptional chemical resistance, and, most important, its ability to adhere to an almost unending variety of materials. Epoxy forms such a tight bond between glass and metal that if you should use epoxy to bond glass into a window frame, the glass will actually crack when contraction from temperature changes takes place.

Technically, epoxy does not use a catalyst, as does a polyester; rather, a curing agent is used which combines with the epoxy. The difference can be most readily seen in its mixing quantities. For polyesters, the ratio of catalyst to resin is about 1 oz. of catalyst to 64 oz. of resin, with the final weight about 64 oz. Epoxies, on the other hand, use a ratio of about 1 quart of curing agent to four quarts of resin, and you will end up with about five quarts of material. Here again, the manufacturer's recommendations for mixing and for the cure time are the *only* guide lines you should follow.

Epoxies may be used for almost any application where polyesters can be used and they have good shelf life. Unmixed, they will keep for almost a year at 72 degrees F., but once mixed, they have a very short pot life.

The solvents used for cleaning brushes and working tools for polyesters are not suitable for use with epoxies. Di-acetone alcohol may be used for cleaning brushes and tools, if they are cleaned before the resin has hardened.

Like other members of the plastic family, epoxies are toxic, and great care should be taken not to let them splash in your eyes or remain on your skin for any length of time.

D.   *Thixotropic Agents*

Some plastic resins are extremely sensitive to temperature changes. At 60 degrees F. they may be as thick as molasses, while at 90 degrees F. they may run like water. And since heat is used as a curing agent, these resins may tend to run off of any vertical or near vertical surface before they have a chance to cure. In order to eliminate this problem, we can add a thixotropic agent.

Among the thixotropic agents familiar in aviation maintenance are micro-balloons. These hollow glass or phenolic balls range in diameter from about ten to three hundred microns and resemble fine sand. If closely examined under a microscope, each micro-balloon is seen to be a perfect sphere. In applying this agent, epoxy or polyester resins are mixed as directed, and the micro-balloons are gently folded in, using care not to beat or crush them. A paste of light consistency is made up and trowled onto the surface where it cures into a hard, light-weight filler. This can be filed or sanded to the required contour.

QUESTIONS

13.   Name four advantages of epoxy resin for use in bonded aircraft structure.

14.   What is the purpose of a thixotropic agent in a plastic resin?

SECTION III:

Laminated Construction

A. Fiberglass Lay-up

Now that we have seen some of the different types of materials used, let's look at the different forms of structure that can be made with these materials.

The first and most generally used bonded structure in general aviation aircraft is a fiberglass lay-up. The laminated sheets can be made either flat or in almost any curved form. And all of these configurations, even flat sheets, require a mold or form.

Flat sheets are laid up by first putting a heavy coat of polyester resin on a glass or aluminum plate. Then a sheet of fiberglass is laid onto the resin. Using a roller, work the fiberglass into the resin, and force all the air bubbles out, then put another layer of resin on top of this, and a second sheet of glass cloth. Continue this procedure until you have the desired number of layers and have worked out all the air bubbles. A final coat of polyester resin is laid over the whole area and the complete stack-up put into a vacuum bag, so all of the air can be drawn out from around the fiberglass and from the resin.

Most manufacturers place their laminated stack-ups in a heater and cure it under both high temperature and pressure. Curved surfaces may be handled in a like manner. The glass and resin are laid up in the same way, the vacuum bag evacuated, and the stack-up put into the oven for curing. In cases where an absolutely smooth surface is required, the mold or model may be first coated with epoxy resin and allowed to cure. Then the fiberglass laid-up over this and cured. The critical side of the finished product determines the type of mold to use. Wing tips, for instance, with their smooth outer surface, are generally cast in a female mold. Here the inside of the mold is liberally waxed and a layer of epoxy resin placed over the entire mold and allowed to cure. After this, the polyester and fiberglass of the specified number of layers is laid up inside the mold. When the complete unit has cured, the part is removed from the mold. It is important to remember that a wax or silicone parting, or release, agent must be used on any mold, either male or female. Without the proper parting agent, the component may become a permanent part of the mold. If the part has contours which will not allow molding in one piece, separate pieces or halves may be molded and joined together with an epoxy joint.

QUESTIONS

15. Why are laminated fiberglass stack-ups placed in a vacuum bag?

16. What determines whether a male or female mold should be used for making a fiberglass part?

17. What is the purpose of a parting agent?

B. Honeycomb

Aircraft structure requires not only strength, but rigidity as well, and the very thin aluminum skins normally used to provide the required strength are lacking in rigidity; so research has produced the honeycomb structure. In honeycomb material, a core of metal, paper, or fiberglass is formed into a cellular structure, and face sheets of fiberglass or aluminum alloy are bonded to either side.

Fig. 3.1 Honeycomb material with a cellular core bonded between two face sheets produces one of the most lightweight, rigid structural materials in use today.

Complex shapes may be constructed of aluminum alloy honeycomb, faced with aluminum alloy sheets, such as the Boeing 747 fore flap,
Fig. 3.2
or flat sheets used for cargo compartment flooring,
Fig. 3.3
or galleys for large transport aircraft.
Fig. 3.4
Smaller aircraft are not without uses of honeycomb. The popular Grumman-American light aircraft uses honeycomb for the fuselage panels and some of the smaller structural components.
Fig. 3.5

Definitely not limited to general aviation aircraft or those used by the airlines, honeycomb construction was actually pioneered for high-speed military airplanes and missiles. Stainless steel honeycomb, furnace-brazed between either stainless steel or titanium skins, is extensively used in some of the more exotic aircraft in the military roster.

Fig. 3.6

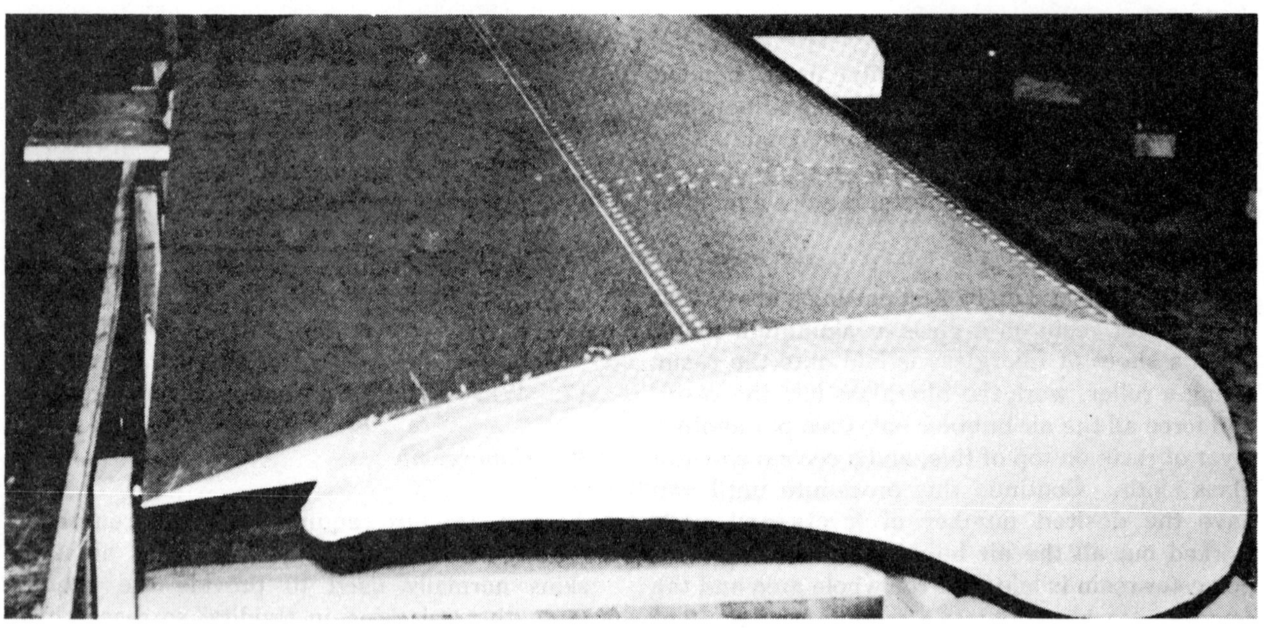

Fig. 3.2   Contoured honeycomb bonded between face sheets of aluminum alloy can be used to make up complex aerodynamic structures.

Fig. 3.3   Flat sheets of honeycomb or end-grain balsa wood between sheets of aluminum alloy make ideal floor and wall panels for modern transport airplanes.

Fig. 3.4 *Galleys in modern airliners are made of honeycomb panels because of their light weight and their rigidity.*

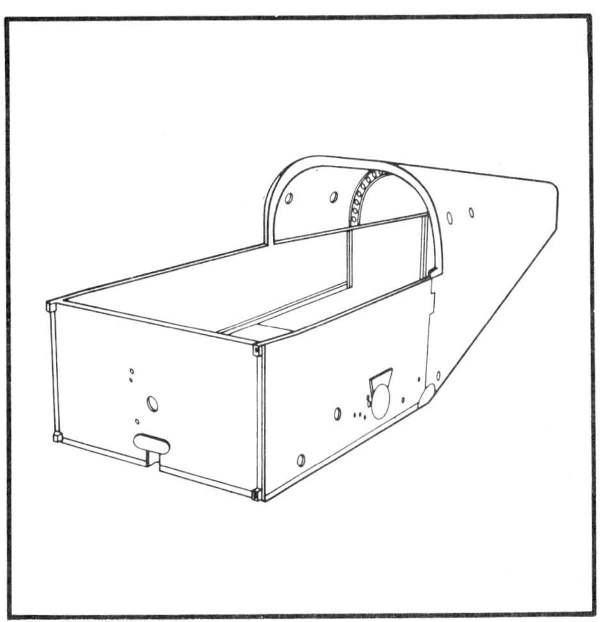

Fig. 3.5 *Bonded honeycomb panels have made the construction of the Grumman-American series of aircraft extremely strong for their weight.*

Fig. 3.6 *Stainless steel honeycomb, furnace-brazed between face sheets of stainless steel, gives aircraft such as this one, great strength for its weight, and its skin resists degradation of strength from the heat generated by the high airspeeds.*

C. *Balsa Core Sandwich*

Balsa wood is familiar to A&P technicians as the prime structural material used by many of us in our first taste of aircraft construction: model airplanes made of balsa and tissue paper. But it is good for high-speed aircraft as well as for model construction.

For sandwich construction of such components as structural panels, floorboards, and many large flat surfaces where rigidity is essential, balsa is sliced across the grain and bonded between two sheets of aluminum alloy or fiberglass. This allows the balsa to be used in compression, a condition where it operates to advantage. End-grain balsa between two face sheets of metal is similar in strength to paper or plastic honeycomb using similar face sheets.

QUESTIONS

18. What is the advantage of aluminum alloy honeycomb over a sheet of plain aluminum alloy for structural panels?

19. What is used to bond stainless steel face sheets to stainless steel honeycomb?

20. In what form is balsa wood used as a structural material in modern aircraft?

SECTION IV:

Bonded Structure Repair

A. Assessment of Damage

Before starting any repair, a complete and total assessment must be made of the damage involved. Since bonded structure has a different nature than riveted construction, it must be inspected in a different manner. However, as with any type of construction, the most important tools an A&P technician uses for his inspection are his flashlight and his eyes.

If there is surface indication of damage on a fiberglass honeycomb or fiberglass laminated sheet, remove the paint from both sides of the structure in the suspect area, and hold a strong light so that it will shine through. In this way you can visually check the extent of the damage. If the honeycomb structure is faced with metal, light will not penetrate it, but areas of delamination may be found by tapping on the surface. Use a coin, such as a quarter, and gently tap the area suspected of damage.

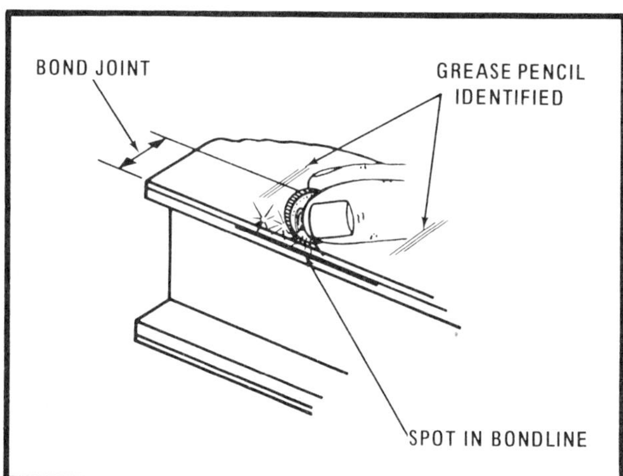

Fig. 4.1   Failed bonds in a laminated structure may be found by tapping the area with a coin. Good bonds produce a ringing sound and delaminations change the ringing into a dull thud.

Delamination usually extends well beyond any area of visible damage so tap below all of the bond lines, and listen for a change of the tone from a ringing sound to a dull thud. Once you have determined the existence of damage and its extent, the type of repair must be decided upon.

On nonstructural components such as fairings, the primary consideration is that we do not make any change in weight or aerodynamic characteristics. But if the part is structural, we must be sure that we preserve the continuity of both the core and the face sheets and retain the strength, weight, and airflow characteristics of the original structure. And, when repairing a radome, we also must take care not to distort its electrical properties. No repair should have abrupt changes in its cross section, as these changes produce stress buildups. Stresses, as you know, flow through the face sheet of a material, through the core to the opposite face, so any repair must transmit these stresses as readily as the original material.

B.   Criteria of a Good Repair

When making a repair to a bonded structure we must use the proper type of material, and proper preparation must be given to the damaged area as well as to the patch itself. Once the patch has been put in place, sufficient pressure and heat must be applied until the cure is effected.

When metal is used as a facing for laminated structure, we must guard against corrosion. Proper protection involves the use of a corrosion-inhibiting primer and a finish coat that completely seals the atmosphere away from the repaired area.

QUESTIONS

21.   How can tapping on a laminated structure indicate an area of delamination?

22.   What happens in the repair of a bonded structure if there is an abrupt change in a cross-sectional area of the structure?

23.   How can corrosion be prevented from getting into the repair of a metal honeycomb structure?

Repairing bonded structure, like most jobs, requires special tools, some of them different from those used in conventional aircraft repair. We must have mixing and measuring equipment for the resins. Polyester resins require the most accurate measurement, which may be done either by volume or by weight, depending upon the manufacturer's recommendations. *When mixing any resin, be sure to adhere to these recommendations in detail.*

Avoid using cups made of plastic or paper to hold the polyester resins, as some may dissolve and contaminate the resins; Pyrex glass measuring cups are best. Remember, when mixing resins for a repair, mix enough extra to use for a test sample.

Fig. 4.2 *When mixing resins for use with bonded structure, be sure that all of the measurements are accurate, and the type of measurement used is that recommended by the resin manufacturer.*

Since epoxies have very little shrinkage when they cure, they are difficult to remove from the mixing cups. For this reason, a mixing plate or pallet is usually used. Most automobile shops use epoxy fillers, and automotive supply houses normally have a good assortment of special mixing tools. These tools are usually made of Teflon or polyvinylchloride, both good release agents, making the tools easy to clean. Mixing pallets made of these materials are also popular and easy to use and clean.

After the resins have been mixed, applied, and allowed to cure, they must normally be smoothed down. This is best done with an air-driven disc or orbital sander.

Air-driven routers are specially suited for repairing bonded honeycomb structure. Hand-held routers spin the cutter at about ten to twenty thousand revolutions per minute, and a collar and support bracket allow the cutter to be raised or lowered so that it can cut through the core without damaging the opposite face material.

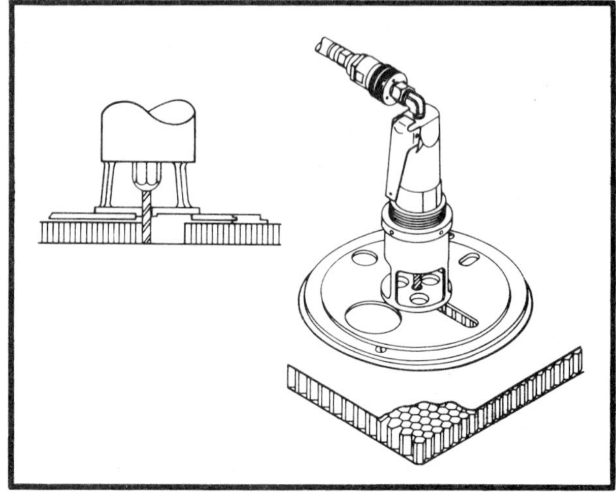

Fig. 4.3 *High-speed routers are used to remove damaged honeycomb core from a panel.*

Heat lamps may be used to accelerate the cure of the resins, and a lamp stand which allows control of the position of the light will make the repair much easier.

One of the less common tools that will assist in making a bonded structure repair is a vacuum pump, one capable of pulling a very low pressure. The vacuum pump like that used to evacuate an air conditioning system is ideal for this, but if such a pump is not available, a compressor salvaged from a commercial refrigerator will normally suffice. These compressors are usually hermetically sealed, with the motor, compressor, and lubricant all in the same unit. For a vacuum bag, heavy polyethylene sheeting available in lumberyards and hardware stores may be used.

After the repair has been made, lay the polyethylene sheet over it and secure the edges with the familiar gray duct tape. Slip the end of the hose from the vacuum pump under the sheeting and seal it tight. When the vacuum pump is turned on, all the air beneath the sheeting will be removed, as well as any air that has been trapped in the resin, and the atmospheric pressure will force the laminations together as they cure. Heat lamps placed near the blanket will accelerate the cure, but you must be careful that the lights are not placed closed enough to melt the sheeting.

Any time we work with chemicals such as resins, accelerators, and catalysts, we should use eye protection; preferably the type of shield that covers the entire face. Water *must* be available, so that in case any of the chemical gets in the eye, we can *immediately* wash it out.

Some form of respirator is necessary, also, during the process of grinding the surface of fiberglass-reinforced plastics, as the tiny pieces of fiberglass can enter the lungs and cause severe damage.

QUESTION

24. What kind of mixing and application tools are best for use with epoxy resins?

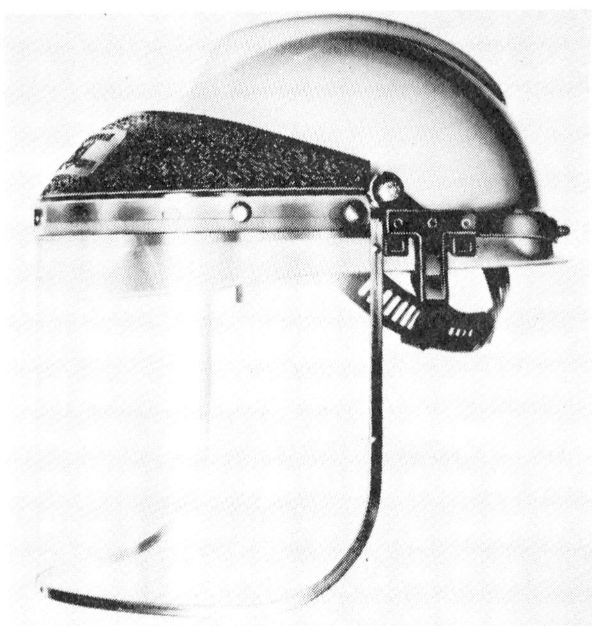

Fig. 4.4   A face mask should be worn to protect your eyes anytime you are working with resins, accelerators or adhesives.

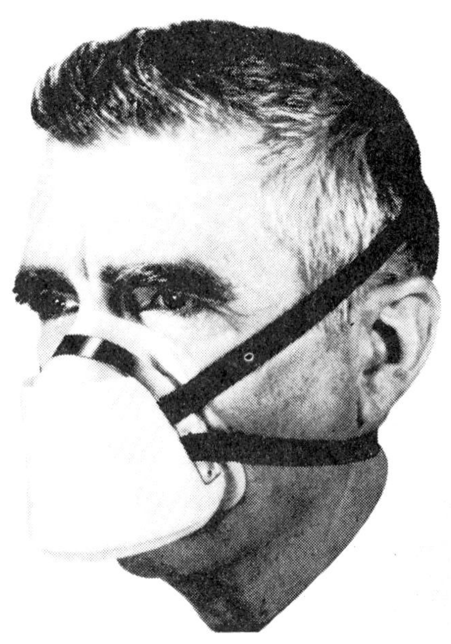

Fig. 4.5   Respirators should be worn any time you are grinding or sanding fiberglass reinforced plastics.

SECTION V:
Specific Repairs to Bonded Structures

A. Damage to Fiberglass Laminated Structure

1. Surface scratches

When the surface of a fiberglass laminated structure is scratched, pitted, or eroded, it should first be washed with detergent and water to remove all of the dirt, wax, or oxide film, and then scrubbed with MEK. After a thorough cleaning, sand it with 280 grit sandpaper, and again use MEK to remove any sanding residue or moisture. This is very essential, as moisture remaining on the surface will inhibit the cure of the resin. Mix enough resin, according to the manufacturer's recommendations, to completely cover the damaged area, and apply one or two coats. Then cover the resin with cellophane or polyvinyl alcohol film, allowing enough overlap to exclude all air from the resin while it is curing and provide a smooth surface for the resin. Tape the film down and rub the surface with your fingers or a rubber squeegee, to work all of the air bubbles out of the resin. After the resin has cured, remove the film and file or sand the surface to conform to the original shape of the part. Refinish it to match the rest of the structure.

2 Delamination

a. Scarf method

If the damaged area is less than about three inches in diameter, the damage may be removed either by sanding with a power sander or hand-sanding with 180 grit sandpaper.

Scarf the edges of the hole back, about fifty times the thickness of the face ply, and thoroughly clean out all of the sanding residue with a cloth wet with MEK.

Prepare the patches by laying the proper weight fiberglass cloth, impregnated with resin (a weight of resin equal to the weight of the patch gives about a fifty-percent ratio, which is correct), on a piece of cellophane or polyvinyl alcohol film. This cloth should be larger than the required patch. Put a second layer of film over the patch and cut it to size. Sandwiching the fiberglass in this way prevents the patch from raveling.

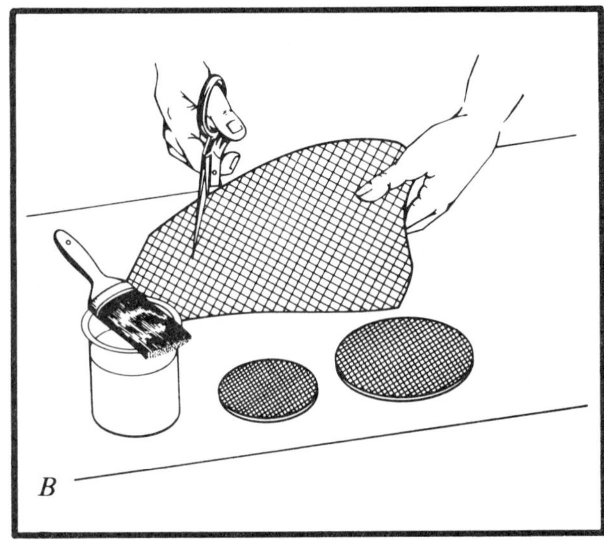

Fig. 5.1   Preparation of fiberglass patches:

A.   Lay the fiberglass cloth on a piece of cellophane and saturate it with the resin.

B.   Lay a second piece of cellophane over the patch and trim the sandwich to size. This makes the handling of the patch easier, and the cellophane is removed as the patch is put into place.

The largest patch should be the size and shape of the outside of the scarfed area, and each of the smaller patches should be the size of the scarfed area of each of the underlying plies—but at least one-half inch smaller than the next larger one. Brush a good coat of resin over the scarfed area and lay in the smaller patch. Work all the air out of the resin and patch, brush on another coat of resin, and lay the next larger impregnated cloth patch over the one just installed, working all the air bubbles out.

Continue with this process until all of the patches are in place. Next, lay a piece of cellophane or polyvinyl alcohol film over the entire repair and carefully work out all of the air bubbles from the resin. Apply pressure over the repair with tape or sandbags, and allow it to cure. After the resin has cured, remove the excess by filing or sanding the repair to the contour of the original part. Smooth the surface with a final sanding, and refinish it to match the original part.

An alternate lay-up method places the larger patch over the area first, and each subsequent smaller patch over this, as is seen in Fig. 5.2.

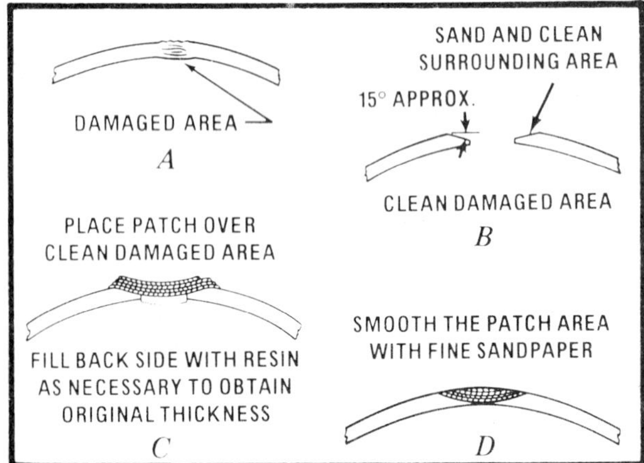

Fig. 5.2   *Repair of laminated fiberglass:*

    A.   *Delaminated or damaged area.*

    B.   *Remove the damaged material and scarf the edges of the hole about 15 degrees.*

    C.   *Lay in sufficient patches to bring the strength back to that of the original structure.*

    D.   *Finish the repair by sanding the new fiberglass to match the original contour.*

b.   Step-joint method

The scarf method of repairing a laminated fiberglass structure or face sheet of a honeycomb structure is the easiest to use, but it does not have the strength of the step-joint type of repair. With a step-joint repair, the damage is outlined with a compass—or, if a square or rectangular repair would be more appropriate, with a straightedge, making appropriately rounded corners. (Use the configuration that will remove the least amount of sound material.) Extend the cleaned-out area for a distance equal to the number of plies to be removed, less one inch. For example, if the repair will necessitate the removal of three plies, the repair will extend two inches beyond the cleared area. Each layer should overlap the one below it by one inch. Use a sharp knife or other type of cutter to carefully cut through the top layer, being careful not to damage the layer underneath.

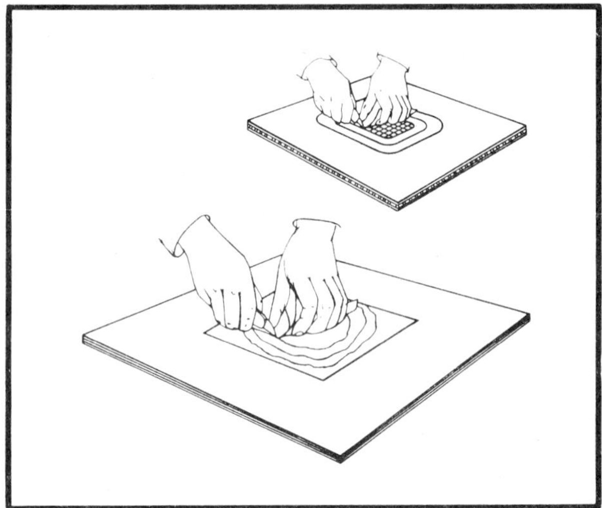

Fig. 5.3   *For a step-joint repair, remove each layer of the original material allowing one inch of each layer to remain.*

Use several passes with the knife, rather than making one deep cut. Begin with one corner of the patch and carefully pry it loose until all of the layer of fiberglass is removed. Next, mark the exposed layer one inch inside of the opening, and carefully cut and remove it. Continue until all of the damaged or delaminated layers have been removed.

Lightly sand, then scrub the entire area with MEK. Prepare the patches as was done for the scarf method, cutting each layer exactly the size of the material removed. Brush in a coat of resin, lay in the patch of the second smallest size, and carefully work out the air bubbles from the resin.

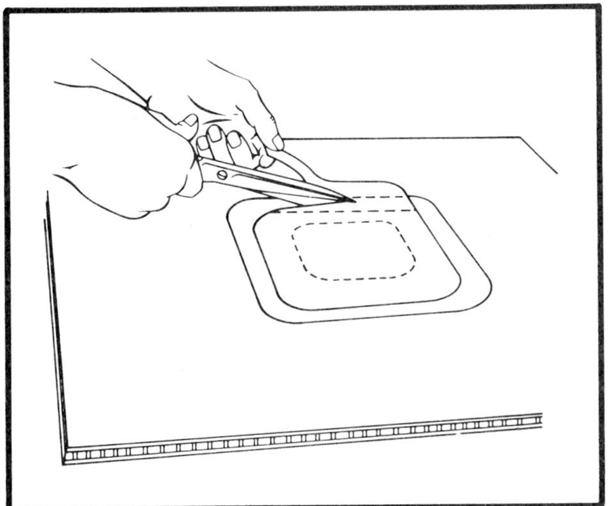

Fig. 5.4   Trim the patches so they will exactly fit into place.

Next, lay in the smaller patches so the layer below will lock it in place.

Butt the top layer to the opening in the face ply, and cover the entire repair with cellophane or polyvinyl alcohol film. Carefully work out all of the air bubbles from the resin and put pressure on the repair with sandbags—or by some other appropriate method.

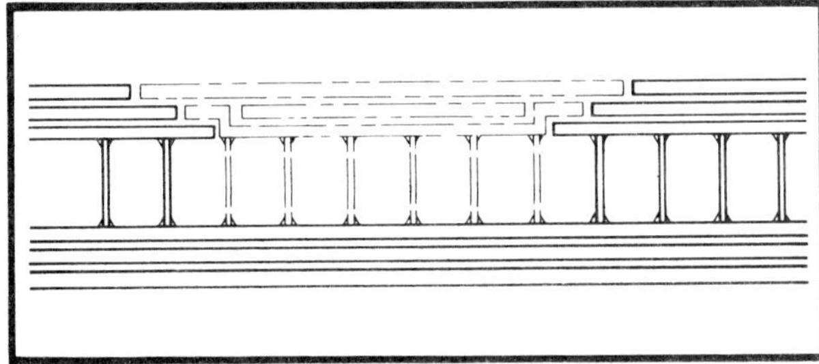

Fig. 5.5   The second ply in a step-cut repair locks the bottom ply in place.

QUESTIONS

25.   What is a good cleaning agent to use when preparing the surface for making a bonded repair?

26.   What can be used to cover a fiberglass-resin repair so that pressure can be applied?

B.   Damage to Honeycomb Structure

1.   Dents

Dents in a honeycomb panel face skin may reduce its yield stress by as much as 20%, but unless strength is critical, small dents are not reworked. For the sake of appearance, however, they may be filled with a standard polyester auto body putty, such as Bondo, sanded smooth, and then painted. This rework, however, does not significantly increase the yield stress of the dented skin.

If it is determined that the face skin must have its original strength, a skin doubler may be applied to the dent, using the same procedure as recommended for a punctured skin.

2.   Surface scratches

These are handled in the same way as scratches in a fiberglass lamination. The surface is cleaned and a coat of resin put on, sanded smooth, then painted to match the rest of the surface.

3.   Surface delaminations

Delaminations of one of the fiberglass skin panels is handled in the same way as described for delaminations of a fiberglass laid-up structure. However, special care must be exercised to protect the honeycomb core from being damaged when the skin is removed. Since the skin is bonded to the core, the use of heat may be required to loosen the bond enough that the face sheet can be removed.

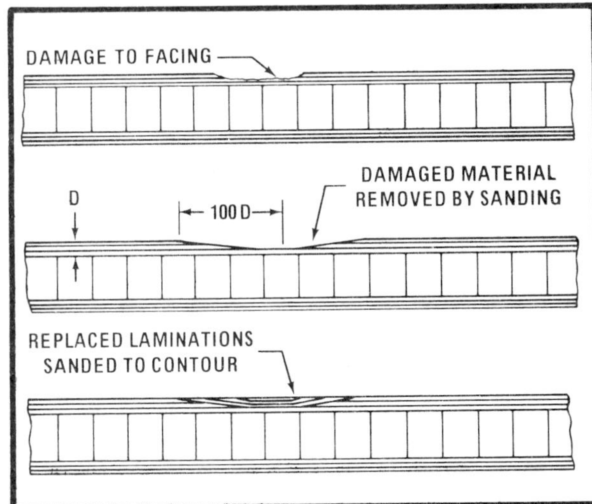

Fig. 5.6   Damaged face plies of a fiberglass skin may be sanded out and replacement laminations bonded in place. After the repair has cured, the surface is sanded to the original contour.

4. Skin penetrated, but core not damaged

If the puncture of a metal face of a honeycomb structure is minor, the repair may be made by potting. First, remove all of the paint for a distance of about two inches around the damage, using fine sandpaper. After the area has been thoroughly cleaned and sanded, scrub it with acetone or MEK, and air-dry it to remove all of the sanding residue. Mix a batch of polyester resin or use a filling compound such as White Diamond, or Bondo, and fill the hole. When it is nearly cured, smooth the top flush with the surface, using a sharp scraper or chisel. It can be sanded to the original contour when the curing process is completed.

Cut a piece of aluminum the same thickness as the original face or one and a half times thicker, and taper it to a ratio of about one hundred to one. This patch should have a minimum radius at its corners of about a half inch. Clean the patch thoroughly and apply a mild acid etch, or roughen it with fine sandpaper. Scrub it with acetone or MEK. Mix a small batch of epoxy resin, and spread it over the damaged area. Center the patch over the hole and cover it with a piece of cellophane. Apply pressure with a sandbag, and allow the repair to cure.

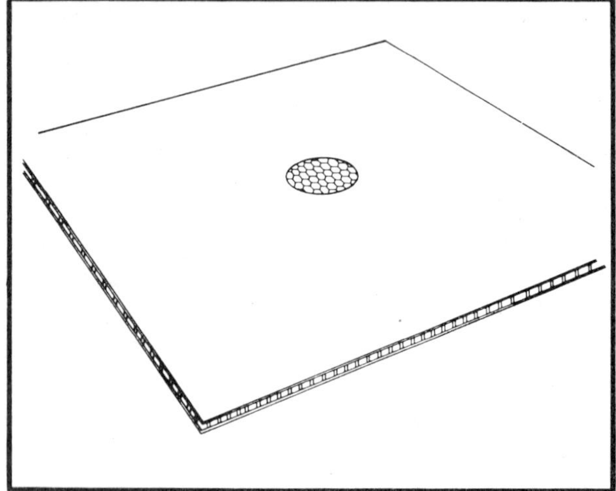

Fig. 5.7 *If the metal skin is punctured, but the damage does not extend into the core, the hole may be filled with filler, and a patch of aluminum alloy bonded in place.*

5. Skin penetrated and core damaged

Potted repairs may be used on areas of damage up to about one inch, and may require removal of some of the face sheet and even some of the core. Remove the face sheet with a power router, using a guide to prevent injury to undamaged face skin. The router may be adjusted to remove the face skin only; the face skin and part of the core; the face skin and all of the core; or both of the face skins and the core.

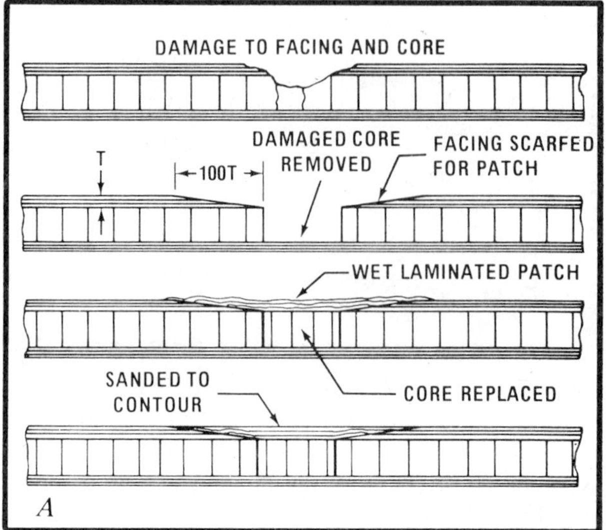

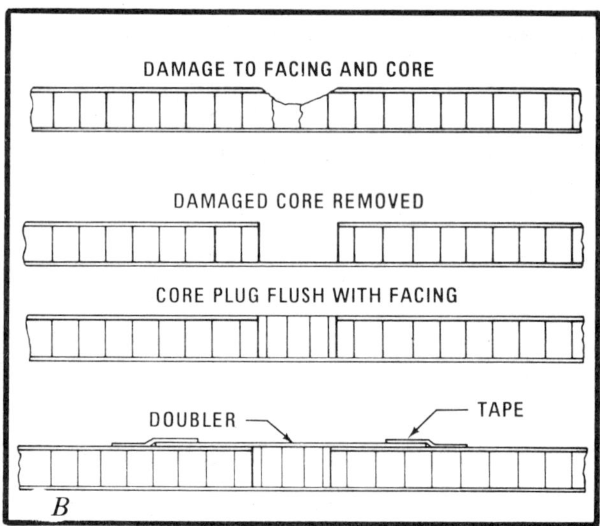

Fig. 5.8 *Facing and core both damaged:*

    A. *Repair to fiberglass skinned honeycomb.*

    B. *Repair to sheet metal skinned honeycomb.*

It may be necessary when a tapered section is being routed, to use wedge-shaped blocks between the routing template and the upper surface. This will allow the router to cut the core material parallel with the lower surface.

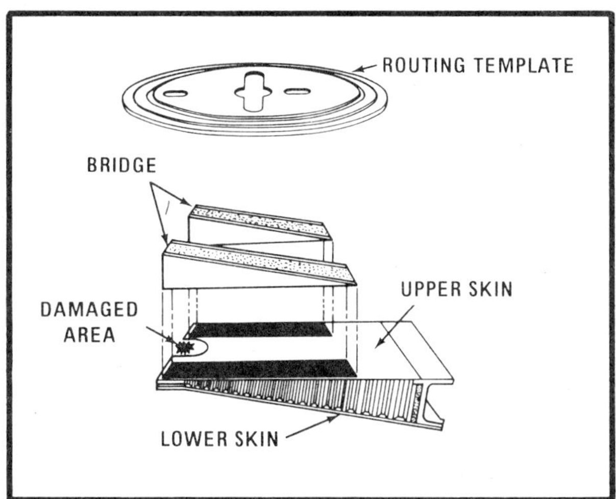

Fig. 5.9   When using a router to remove the core of a tapered honeycomb panel it is necessary to use tapered wedges to hold the routing template parallel with the lower skin.

Mix a sufficient quantity of polyester resin to fill the hole. Micro-balloons may be added to the resin, if desired, to serve as a filler. With the resin and filler thoroughly mixed according to the manufacturer's recommendations, pour the mixture into the hole, filling all of the cells, then work out all of the bubbles with a toothpick. When the cure is nearly complete, use a scraper or a sharp chisel to make a smooth, flat surface. Prepare the patch, using a circular, oval, or rectangular form, with corners having at least a half-inch radius. Taper the edges to provide a smooth junction with the skin, and bond it in place with epoxy, using the appropriate heat and pressure.

If the damaged area is larger than can be repaired by potting, a plug may be cut and bonded in place.

Remove all of the damaged honeycomb core, and prepare either a balsa wood or honeycomb replacement plug. If balsa is used, cut the plug so the grain is perpendicular to the skin. Sand it lightly with 400 grit sandpaper and wipe off all of the dust; then do not touch the plug with your bare hands until it is ready to be installed. If honeycomb material is used, it should be as near the same density as the original as possible, and the insert plug should completely fill the damaged area, with its top edge even with the adjacent skin.

Be sure that all traces of paint and primer are

removed from the skin surrounding the hole, back for about two inches, then roughen the skin with 400 or 600 grit sandpaper. Once again, mix sufficient epoxy resin as recommended by the manufacturer. If a balsa plug is used, spread the adhesive lightly over all of its surfaces. However, if an aluminum honeycomb plug is being used, brush or trowel adhesive into the exposed surfaces of the existing honeycomb and everywhere the plug makes contact with the core; then insert the plug into the hole.

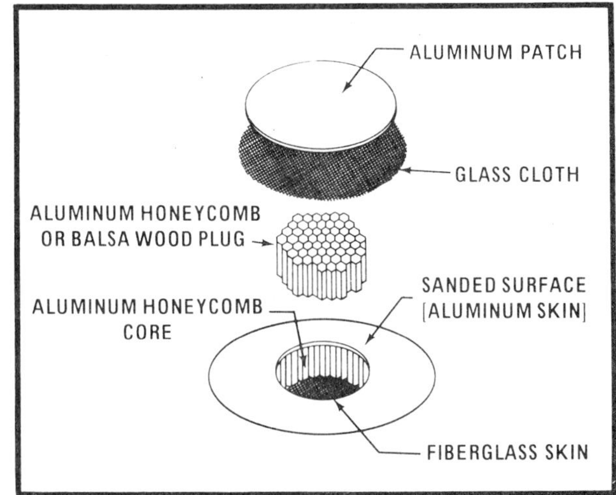

Fig. 5.10   Damaged core material is routed out to the lower skin and a filler of balsa wood or honeycomb is inserted. A replacement skin of glass cloth and aluminum alloy is bonded in place over the repair.

Cut a piece of fiberglass cloth large enough to cover the plug hole by about one-eighth of an inch, impregnate it with the epoxy resin, and lay it in place on the clean surface, centering it over the hole. Prepare a circular patch of 0.012 to 0.015 inch aluminum alloy to cover the hole and extend about one and a half to two inches beyond. Roughen the surface of the patch with fine sandpaper, clean it thoroughly, and center it over the glass cloth. Cover the patch with cellophane or polyvinyl alcohol sheet and apply pressure, making sure that neither the patch nor the glass cloth slips out of place. Remove any excess adhesive with a cloth dampened with MEK or toluol, and cure the patch according to the recommendations of the adhesive manufacturer. After the patch has cured, remove the pressure pads, sand away all of the excessive adhesive, and prime and finish the area to match the surrounding surface.

6. Both skins penetrated and core damaged

Fabricate the core plug, the glass cloth, and the sheet metal patches, as was done for the repair when only one face skin was damaged.

Make a temporary block to hold the plug in place while one side is repaired in the way previously described. After one side has cured, remove the temporary block and repair the remaining side.

7. Riveted repairs to bonded structure

It is sometimes difficult to control the curing of bonding resins. For this reason, Grumman-American Corporation has devised a series of repairs to their bonded structure in which the repairs are attached by rivets instead of adhesive.

If one face skin is damaged less than one inch and there is only superficial damage to the core, the damage may be trimmed, the skin cleaned, and a doubler made of 0.040 2024-T3 aluminum alloy, prepared to extend one inch around the cut-out area.

The exposed honeycomb is sealed with an inhibited sealer and the doubler riveted to the surface with one-eighth-inch Cherry rivets. After the patch is finished, it is faired into the existing skin with epoxy filler.

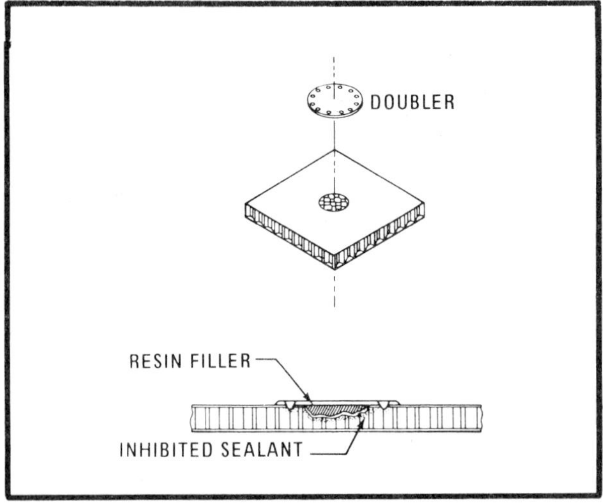

Fig. 5.11   Damage to small areas affecting only one skin and a portion of the core may be repaired by filling the damaged core with an approved sealer and riveting a doubler over the cleaned out hole, using Cherry rivets.

If both skins and the core have been damaged, a doubler of 0.024 2024-T3 aluminum alloy may be riveted in place on both sides, using Cherry rivets. The honeycomb repair section is sealed with inhibited sealer, and all of the rivets are dipped in sealer before they are installed.

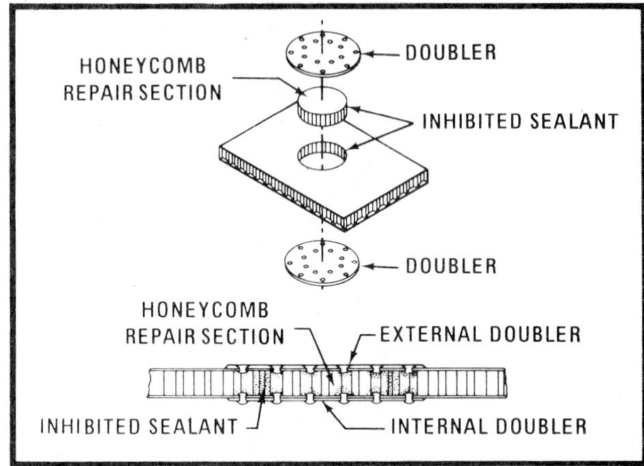

Fig. 5.12   If both skins and the core have been damaged, a repair can be made by replacing the damaged material, sealing all of the edges, and riveting a doubler on both sides of the panel using Cherry rivets.

Larger repair sections may be spliced in place using doublers of appropriate thickness on each side, attached with Cherry rivets. The edges of the honeycomb are sealed with inhibited sealer, and as before, all of the rivets are dipped in sealer before they are installed.

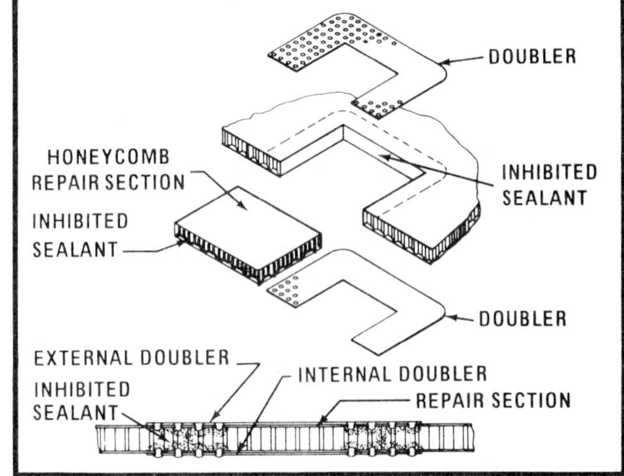

Fig. 5.13   Repair sections are spliced into a damaged panel by sealing all of the edges of both the panel and the repair section, and joining the two parts with doublers on both skins using Cherry rivets.

27. What may be used to fill in a core area in which the core damage is only about one half inch in diameter?

28. What may be used to fill in the area from which a damaged core has been removed, if the damage is larger than can be repaired by potting?

C. Radome Repair

Perhaps the most critical repair of bonded structures is that of a radome, or the housing that covers the radar antenna. This structure must be aerodynamically smooth, since it usually forms the nose of the airplane, structurally strong, as it is exposed to all of the aerodynamic loads of the fuselage, and electrically transparent. This means that any repair made must not distort the reception or transmission of energy from the radar equipment.

If there is any question as to the specifics of a repair, the airframe manufacturer should be consulted before the repair is begun. A typical repair is shown in Fig. 5.15. If the damage, after it is cleaned out, does not exceed half an inch in diameter within the radar window—that is, the area swept by the transmitted energy—or one inch outside the window, it may be repaired as shown. A thin metal plate is separated from the radome with a piece of polyethylene sheet and taped in place. The hole is filled with an approved filler, allowed to cure, and trimmed flush with the surface. A resin-impregnated fiberglass patch is placed over the repair, overlapping it by half an inch. All of the air is worked out of the patch, and it is covered with a sheet of polyethylene, it is worked free of air bubbles and allowed to cure. After one side has cured, the metal plate is removed and replaced with a similar patch. The repair is finished by lightly sanding all of the roughness from the patches.

Radomes should not be painted unless specifically directed in the repair manual, and then using only the paint specified.

Fig. 5.14 A radome is one of the more critical applications of bonded structure. It must be aerodynamically smooth, structurally strong, and electrically transparent.

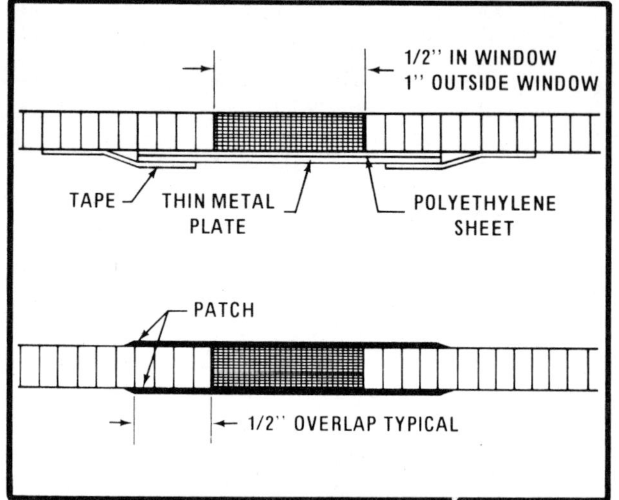

Fig. 5.15   Radomes are repaired by inserting an approved filler, trimmed to the proper contour after it has cured, and a fiberglass patch placed over the repair.   Be sure that the manufacturer's recommendations are followed in minute detail for this type of repair.

D.   General Considerations for Repair to Bonded Structure

Repair of bonded structure, like that for any other type of construction, must be done in accordance with approved data.   This simply means that before any work is started, you must know the methods and materials that are approved for the specific job.   Adhesives, resins, fillers, and other repair materials and methods change so rapidly that it is impossible for a book of this nature to specify brands and numbers, or to dimension repairs.   Service manuals for each airplane list detailed procedures, specifying the resins and other materials approved for the repair.   For repairs not described in the service manual, consult the aircraft manufacturer for the approved method and materials.

1.   Cutting information

Whenever possible, sheet metal and fiberglass laminates should be sheared rather than saw-cut.

Material	Blade Type	Comments
Wood only	All Purpose Wood	Good for most woods. Combination Wood blade necessary for hardwoods.
	Carbide Tip	Can be used but will dull.
Fiberglass only	Diamond Grit	Best.
	Carbide Grit	Does similar job as diamond but not as hard as diamond.
	Carbide Tip	Not recommended. Dulls rapidly.
Fiberglass/Wood	Carbide Tip	Best, although dulling will occur. Grit blade will clog.
Fiberglass/Balsa	Carbide Tip	Best.
	Diamond Grit	Possible tendency for grit to clog.
	Carbide Grit	Ditto.
Fiberglass/Paper Honeycomb	Carbide Tip	Best. Make sure blade is sharp.
	Diamond Grit	Possible, some tendency to clog.
	Carbide Grit	Ditto.
Aluminum/Wood Stainless/Wood	Carbide Tip	All other blades are unsuitable.
Aluminum or Stainless/ Aluminum Honeycomb	Carbide Tip	All other blades are unsuitable.
Fiberglass/Aluminum Honeycomb	Carbide Tip	All other blades are unsuitable.

Fig. 5.16   Saw types recommended for bonded structure repair

Shearing is not only more efficient, but is cleaner, as it minimizes fiberglass dust. When sawing fiberglass, be sure to use a suction system to collect the dust.

Band saws should not be used because of the poor control they afford, but for both fiberglass laminates and sandwich panels circular saws will do a good job.

Fig. 5.16 lists blades that are recommended for cutting bonded structural materials. Be sure that any blade you use is reasonably sharp, as even the recommended blades will not give a satisfactory cut if they are dull.

Fig. 5.17 gives information on saw blades suitable for use on bonded structural materials.

2. Helpful hints for bonding sandwich materials

The M.C. Gill Corp., one of the recognized leaders in bonded structural materials, offers the following hints:

Adhesives join materials primarily by bonding to their surfaces and the major variables in bonding materials are: treatment of the surfaces to be bonded, and the property of the adhesives.

a. Surface treatment

In order to get a good bond, an adhesive must thoroughly wet the surfaces to be bonded. The bondability of a surface is directly related to the surface preparation and the surface roughness.

Thermoset fiberglass: sand the surface with 120-grit dry sandpaper until all of the gloss has been removed, then solvent-wipe with acetone or MEK.

Aluminum: preferably sulfuric-chromic acid etch.

Stainless or carbon steel: degrease or solvent-wipe with acetone or MEK, then immerse in a pickling solution for 10 minutes at 70°F.

Type	Size	Approximate Cost		Manufacturer	Remarks
		New	Regrit or Sharpen		
Diamond Grit	8''	$148.00	$134.00 w/exch.	O'Rourke Diamond Company 1411 Van Owen St. North Hollywood, CA	Std. Slot Blade
	10''	$215.00	$193.00 w/exch.		
Carbide Grit	8''	$ 11.25		Glendale Saw Works 520 W. Colorado Pasadena, CA	Available only 8'' or 10''. No regritting.
	10''	$ 12.42			
Carbide Tip	8''	$ 88.00	$ 16.00 (40 teeth)	A-1 Carbide 10637 Midway Ave. Cerritos, CA	8'' Blade = 40 teeth 10'' Blade = 50-60 teeth
	10''	$102.00	$ 17.05 (50 teeth)		

Blade Diameter In Inches	Revolutions Per Minute	Surface Speed Feet Per Minute
8''	4500	9400
10''	4000	10500
12''	3600	11300
16''	3450	14500

Fig. 5.17 Saw information for cutting bonded structure materials

Pickling solution:

    1-part by weight 37% hydrochloric acid
    2-parts by weight water

Rubber or elastomers (except polyurethane, thiokol, or silicone): scrape with blade, abrade with 240-grit sandpaper, then scrub with toluene-soaked rag. The surface must be completely dry before bonding.

Polyurethane, polysulfone, and other thermoplastics: abrade the surface and solvent-wipe with acetone or MEK.

b. *Property of the adhesive*

The property of the adhesive must be balanced between the materials to be bonded and the environment in which the joint is going to serve. See Fig. 5.18.

3. *Sealing balsa sandwich panels*

Again, from the M.C. Gill Corporation we have information on balsa wood sealers.

Thin epoxy sealant such as Pro-Seal 825 has been proven to be the most effective sealant for balsa panels. However, with only a small sacrifice in sealing properties, a one-part rubber solution, such as Products Research PR1005L, requires no catalysis and is therefore easier for shop workers to apply. Since both of these solutions are thin, with a tendency to run, they are best applied with a paintbrush.

1. Thermoset—for structural bonding and where long term resistance to creep, moisture and fatigue are primary considerations.

To Bond	Type of Bond	Adhesive Type	Adhesive #	Vendor
metal, wood, thermoset plastics	flexible	2-part epoxy	EC2216 A/B	3M Company 6411 Randolph L.A., CA 90040
metal, wood, thermoset plastics	rigid	fire retardant 2-part epoxy	Gillab 2020	M.C. Gill Corp. 4056 Easy Street El Monte, CA 91731
rubber, metals, glass, plastics	structural-rigid	2-part epoxy	1177 A/B	B.F. Goodrich 5701 E. Union Pacific L.A., CA 90022
thermoplastics	flexible	urethane	5738ABX	Furane 5121 San Fernando Rd. West L.A., CA 90039

2. Thermoplastic—contact and pressure-sensitive applications. For non-critical bonding where speed and ease are primary considerations.

To Bond	Type of Bond	Adhesive Type	Adhesive #	Vendor
woods, metal, rubber, vinyl trim	contact-room temp.	neoprene rubber	SE176-9	H.G. Fuller Co. St. Paul, Minn.
woods, metal, rubber, vinyl trim	contact-room temp.	neoprene rubber	Anchorweld 306	Chemware City of Industry, CA
foams	contact-room temp.	elastomer	8140	Swift Chicago, Ill.

Fig. 5.18 Adhesives for bonding sandwich materials

Sealant	Type	Mix Ratio	Cure Time	Pot Life	Viscosity	Application	Relative Price
unsealed	——	——	——	——	——	unsealed	——
PR-1005L	1-part nitrile rubber	1-part no mixing required	30 min.	24 hrs.	thin	quick dry solvents very thin coating, tinted red	$12.27/gal. coverage: 200 sq. ft./gal.
Pro-Seal #825	2-part epoxy	100:100 by volume	2 hrs.	12 hrs.	thin	quick dry solvents, gray color, coating runs easily	$13.25/gal. coverage: 160 sq. ft./gal.
PR-1436	2-part polysulfide	100:10 by weight	16 hrs.	1 hr.	paste	easy to apply, rubber cure, offensive odor	$34.66/gal. coverage: 80 sq. ft./gal.
Pro-Seal #890	2-part polysulfide	100:9 by weight	8 hrs.	2 hrs.	semi-paste	easy to apply, hard rubber cure, slight odor	$34.15/gal. coverage 80 sq. ft./gal.

Fig. 5.19   Sealants for sealing balsa sandwich panels

Polysulfide sealants may be used, but their odor, expense, and slow drying time make them less than a good selection for sealing balsa panels.

QUESTIONS

29. What kind of paint should be used on a radome repair?

30. Why is it recommended that fiberglass laminates be sheared rather than sawed whenever practical?

31. What solvents are recommended for cleaning the surface of a fiberglass laminate *before* you apply the bonding adhesives?

4.   Applying pressure and heat for curing

Small repairs may be cured by inserting a hose from a vacuum pump under a piece of polyethylene sheet and sealing the edges with duct tape. The air is evacuated from under the sheet, and all of the air bubbles are worked out of the resin. A heat lamp may be directed onto the repair, but care should be exercised that there is not enough heat built up on the surface that the polyethylene will be damaged.

In Fig. 5.20 we have a simple method of applying pressure to a patch which covers only the face sheet of the honeycomb. The hole is filled with the appropriate potting compound and the tapered patch is put in place. This is covered with cellophane or a polyvinyl sheet, a cushion of foam rubber is laid over it, and a pressure plate of quarter-inch plywood is screwed through it to the honeycomb, using long sheet metal screws. After the resins have cured, the screws are removed and the holes filled with the appropriate filler and cured. After the curing process is completed, the surface is trimmed of any excess resin and painted to match the rest of the structure.

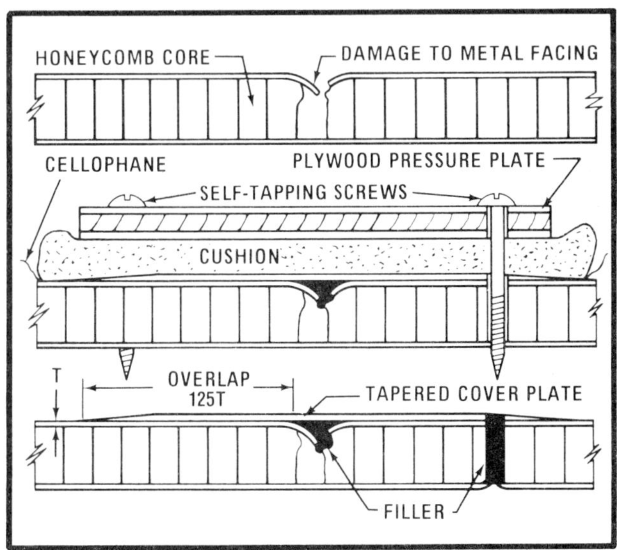

Fig. 5.20   Cushion method of applying pressure to the face skin of a metal skin honeycomb structure

Fig. 5.21 shows the same type of pressure application for a repair that involves the installation of a back-up plate on the inside of the structure, a core replacement, and a tapered patch on the top surface.

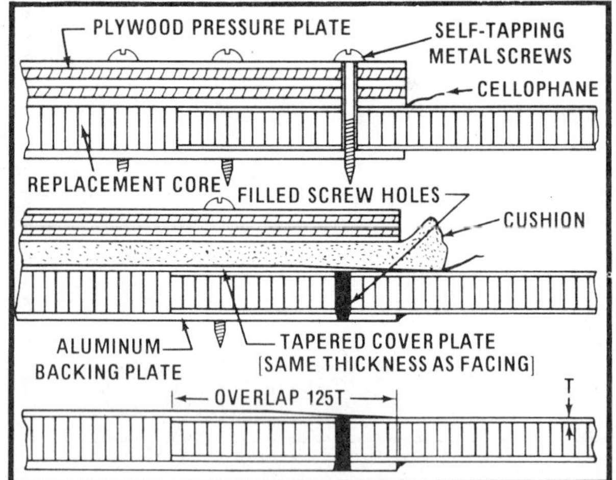

Fig. 5.21   Cushion method of applying pressure to a honeycomb repair using a backing plate on the inside of the structure

Some large, high-speed aircraft require bonded structure repairs to be made in areas where it is very difficult to apply pressure. In this situation an inflatable rubber diaphragm pressure plate may be used. These are manufactured by Airline Systems of San Carlos, California. The patch is put in place and the pressure plate is placed over it and anchored with suction cups. The rubber diaphragm is inflated with a hand pump to the pressure specified for the repair.

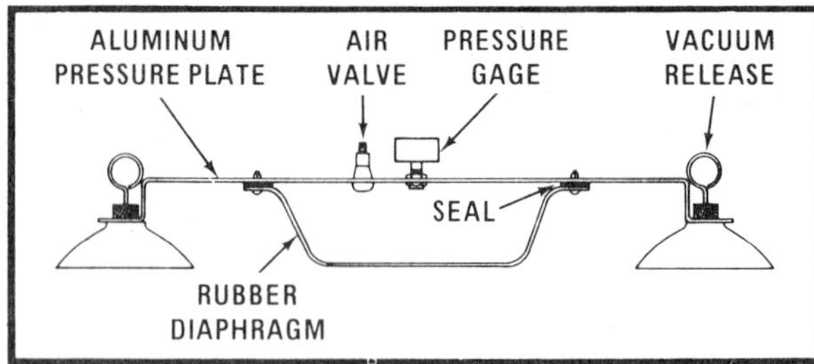

Fig. 5.22   Airline Systems' rubber diaphragm pressure plate method of applying pressure for making repairs in difficult locations

If heat is required to accelerate the cure, a chemical heat pack, also made by Airline Systems, may be used in conjunction with the air bag type of pressure plate. The heat pack is put in place and water supplied to it as shown in Fig. 5.23. Any vapors that are released are vented through a vapor trap so that there will be no buildup to vary the pressure applied.

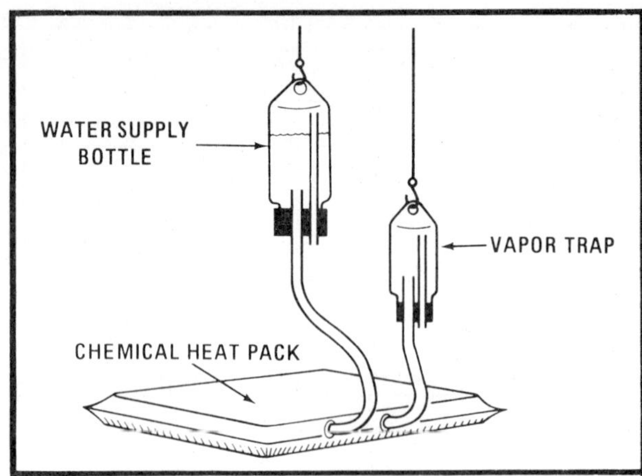

Fig. 5.23   Airline Systems' chemical heat pack for providing heat to accelerate the cure of a bonded structure repair

SUMMARY

In summary, a few basic considerations of bonded structure repair are well worth restating:

1.   In all details, follow the recommendations of the manufacturer of the materials being used. Any attempt to improve on their methods can only lead to problems, both technical and legal.

2.   Corrosion is one of the main enemies of metal bonded structure, and any repair should be completely sealed to prevent the entry of any moisture.

3.   Dents can be filled to give an aesthetic appearance to a surface, but remember that strength can only be replaced by added structure, such as doublers, bonded or riveted over the filled dent.

4.   Cure time of many adhesives may be hastened by the application of heat; but be careful that the heat is not too intense and that it is spread out enough that no local hot spots are created.

QUESTIONS

32.   What precautions should be taken when using heat lamps to accelerate the cure of a bonded structure repair?

33.   How can corrosion be prevented in a bonded structure repair?

34.   How can the strength be restored to a metal-faced honeycomb panel that has been dented?

SECTION VI:
Transparent Thermoplastic Material

A. Types

1. Cellulose acetate

Two types of transparent thermoplastic material used for aircraft windows and windshields are those with a cellulose acetate base, and acrylics. The cellulose acetates were used in the past, but because of their inferior qualities have just about passed from the scene. They are not considered suitable substitutes for acrylics.

2. Acrylics

Acrylic plastics recognized by the trade names of Lucite or Plexiglas, meet the military specifications of MIL-P-6886 for regular acrylic plastic, MIL-P-5425 for heat-resistant acrylic plastic, and MIL-P-8184 for craze-resistant acrylic.

One way to tell if the plastic is acrylic or acetate is by rubbing a bit of acetone on a small piece of the material in question. If it turns white but does not soften, it is acrylic, but if it softens and does not change color, it is most probably acetate. A final check can be made by burning a scrap of the material. Acrylic burns with a steady, clear flame, while acetate burns with a sputtering flame and dark smoke. After the flame is extinguished, the odor left by burning acrylic is fairly pleasant, but that from acetate is definitely *unpleasant*.

Fig. 6.1   The windshields of modern, general aviation aircraft are made of acrylic pastic.   Acrylics produce windshields that are lightweight, strong, and have excellent optical qualitites.

B. Storage and Handling

As transparent thermoplastics will soften and deform when they are heated, they should be stored in areas where the temperatures are not excessive. Plastic sheets should be kept in a cool, dry location, away from heating coils, radiators, hot water or steam pipes. They should also be kept away from such fumes as may be found in a paint spray or paint storage area.

Paper-masked transparent plastic sheets should be kept indoors, as direct rays of the sun will accelerate the deterioration of the adhesive, causing it to cling to the plastic and making removal of the masking paper difficult.

These plastic sheets should be stored with the masking paper in place and in bins which are tilted at approximately a 10 degree angle from the vertical. This will prevent buckling. If it is necessary to store sheets horizontally, care should be taken to avoid getting dirt and chips between the sheets. Stacks should not be over 18 inches high, with the smaller sheets stacked on the larger ones to avoid unsupported overhang. Masking paper should be left on the plastic sheets as long as possible and care should be exercised to avoid scratches and gouges caused by sliding sheets against each other or across rough or dirty tables.

Formed sections should be stored with ample support to lessen their tendency to lose shape; vertical nesting should be avoided. Protect formed parts from temperatures higher than 120°F (49°C). Leave their protective coating in place until they are installed on the airplane, as this will help prevent scratches.

QUESTIONS

35. What type of plastic is used for most aircraft windshields?

36. What is the easiest way to tell acrylic from acetate plastic?

37. What precaution should be observed when storing acrylic sheets vertically?

C. Forming Procedures and Techniques

Transparent acrylic plastics become soft and pliable when they are heated to their respective forming temperatures, enabling them to be formed to almost any shape. On cooling, the material retains the shape to which it was formed, except for a small amount of contraction. It is not desirable to cold-form compound curvatures with transparent acrylic plastics, or to spring them into a curved frame without heating.

Practical Oven Temperatures								
Thickness of sheet (in.)	0.125		0.250		0.125		0.250	
Type of forming	Regular acrylic plastic, MIL-P-6886				Heat-resistant acrylic plastic, MIL-P-5425, and craze-resistant acrylic plastic, MIL-P-8184			
	°C.	°F.	°C.	°F.	°C.	°F.	°C.	°F.
Two-dimensional (drape)	113	235	110	230	135	275	135	275
Stretch forming (dry mold cover)	140	284	135	275	160	320	150	302
Male and female forming	140	284	135	275	180	356	170	338
Vacuum forming without form	140	284	135	275	150	302	145	293
Vacuum forming with female form	145	293	140	284	180	356	170	338

Fig. 6.2 Recommended temperatures for forming acrylic plastics.

Acrylic plastic may be cold-bent into a single curvature if the material is thin and the radius around which it is bent is at least 180 times the thickness of the sheet. Cold-bending beyond these limits causes great stress to be imposed upon the surface of the plastic. This may result in tiny fissures, called crazing.

1. Heating

Prior to heating any transparent plastic material, remove all of the masking paper and adhesive from the sheet. If the sheet is dusty or dirty, it should be washed with clean water and rinsed well. Dry it thoroughly by blotting with soft absorbent paper.

For the best results in hot-forming acrylics, use the temperature recommended by the manufacturer. The table in Fig. 6.2 is typical for MIL-specification acrylics. A forced-air oven should be used, one capable of operating over a range of 120° to 374°F (49° to 190°C).

If the recommended temperature is exceeded during the forming of acrylic plastics, bubbling on the surface may occur and impair the optical qualities of the sheet.

For uniform heating it is best to hang the sheets vertically. This can be done by grasping the sheets by their edges with spring clips such as shown in Fig. 6.3.

If the piece is too small to be held by clips, or if there is not enough trim area, the sheets may be laid on shelves or racks upon soft felt or flannel.

Be sure that there is enough open space to allow the air to circulate around the sheet and heat it uniformly.

When handling the plastic, it is wise to wear cotton gloves, as this will prevent finger marks or other damage on the soft surface.

For small forming jobs, such as landing light covers, a kitchen baking oven may be used for heat. Infrared heat lamps may also be used, provided they are arranged on seven- or eight-inch centers with enough of them in a bank to heat the largest size sheet evenly. To get the most uniform distribution of the heat, the lamps should be placed about 18 inches from the material.

Never use hot water or steam directly on the plastic to heat it, as this would likely cause the acrylic to become milky or cloudy.

2. Forms

Heated acrylic plastic will mold with almost no pressure, so the forms used can be of very simple construction. Forms made of pressed wood, plywood, or plaster are sufficient to mold simple curves, but reinforced plastic or plaster may be required to shape complex or compound curves.

Since plastic will conform to any waviness or unevenness, the form used should be completely smooth. To ensure this, the form should be sanded and covered with soft cloth such as outing flannel or billiard felt.

The mold should be large enough to extend beyond the trim line, and provisions must be made for holding the hot plastic against the mold while it cools.

A mold can be made for a rather complex part by using the damaged part itself. If the part is broken, tape the pieces together, wax or grease the inside so the plaster will not stick to it, and support the entire part in sand. Fill the part with plaster, allow it to harden, and remove it from the mold. Smooth out any rough parts, cover it with soft cloth and it is ready to be used to form the new part.

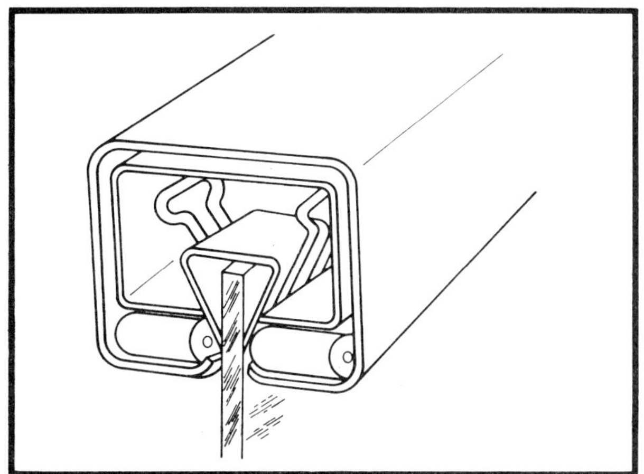

Fig. 6.3 Spring clips may be used to hold acrylic sheets vertical while they are being heated.

3. Forming methods

a. Simple curve forming

Heat the plastic material as we have just described, remove it from the oven, and carefully drape it over the prepared form. Carefully mold the hot plastic to the form. Either hold or clamp the sheet in place until it is thoroughly cooled. This process may take from ten minutes to half an hour, but do not force-cool it.

b. Compound curve forming

This type of forming is normally done for such parts as canopies or complex wing tip light covers, and it requires a great deal of specialized equipment. There are four commonly used methods, with each having its advantages and its disadvantages.

[1] Stretch forming

Preheated acrylic sheets are stretched mechanically over the form in much the same way as was done with the simple curved piece. Special care must be taken to preserve uniform thickness of the material, as some parts will have to stretch more than others.

[2] Male and female forming

This requires expensive, matching male and female dies. The heated plastic sheet is placed between the dies, which are then mated. When the plastic cools, they are opened.

[3] Vacuum-forming without forms

Many aircraft canopies are formed in this way. A clamp with an opening of the desired shape is placed over a vacuum box to hold the heated sheet in place. When the air in the box is evacuated, the outside air pressure will force the hot plastic through the opening, forming a concave bubble. The surface tension of the plastic will shape the bubble.

[4] Vacuum-forming with female form

If the shape desired is other than that which would be provided by the surface tension, a female mold, or form, is used. It is placed below the plastic sheet and the vacuum pump connected. When the form is evacuated, the outside air pressure will force the hot plastic sheet into the mold and fill it.

QUESTIONS

38. What is likely to happen to a piece of acrylic plastic that is cold-formed over too small a radius?

39. Why is it not a good idea to heat transparent plastic in hot water or steam?

40. To what temperature should a piece of one-quarter-inch thick acrylic be heated to form it over a two-dimensional drape form?

4. Sawing and drilling

a. Sawing

Several types of saws are suitable for use with transparent plastics; however, circular saws are the best for straight cuts. The blades should be hollow-ground or have set, or swaged, teeth to prevent binding. After they are set, the teeth should be side-dressed on the machine to produce a smooth edge on the cut.

In order to prevent the acrylic sheet from becoming overheated, it should be fed into the saw slowly. If the plastic begins to smoke or the edges of the cut appear smeared, it may be an indication that the feed has been too fast for the thickness of the material or for the type or condition of the saw.

Band saws are recommended for cutting acrylic sheets when the cuts must be curved or where the sheet is cut to a rough dimension to be trimmed later. Close control of size and shape may be had by band-sawing a piece to within 1/16'' of the desired size, as marked by a scribed line on the plastic, and then sanding it to the correct size with a drum or belt sander.

b. Drilling

Unlike soft metal, acrylic plastic is a very poor conductor of heat, and when it is drilled, provisions must be made for removing the heat. Deep holes require cooling, and it has been found that a water-soluble cutting oil is a satisfactory coolant, as it has no tendency to attack the plastic.

The drill used should be carefully ground and free from nicks or burrs which would affect the surface finish. The drill should be ground with a greater included angle, Fig. 6.4, than would be used with soft metal. The rake angle should be zero.

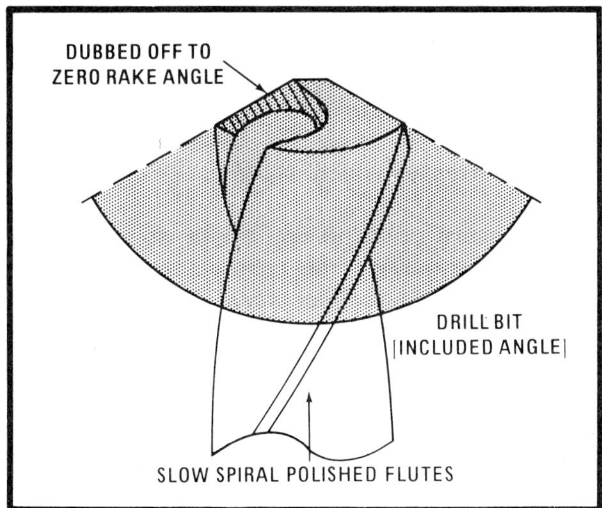

Fig. 6.4   The drill used for acrylic plastics should have smooth flutes, zero rake angle, and the included angle should be greater than that used for drilling aluminum.

For drilling small holes in aircraft windshields and windows, the patented Unibit, a stepped drill, may be used. It is able to cut holes from 1/8'' to 1/2'' in 1/32'' increments and it produces good smooth holes with no stress cracks around their edges.

QUESTIONS

41.   What should be done to the teeth of a circular saw blade used to cut acrylic sheet, *after* they have been set?

42.   What is indicated if the edges of a saw cut in an acrylic sheet appear to smear?

43.   What is a good coolant to use when drilling acrylic plastic?

44.   Would a properly sharpened drill for use with acrylics, have a greater or smaller tip included angle than one used for drilling aluminum?

5.   Cementing

a.   Application of cement

Acrylic plastics may be joined by using ethylene dichloride, a clear liquid cement. It softens the material and forms a cushion between the two pieces, allowing a thorough intermingling of the surfaces.

There are two methods of cementing plastics; soaking or gluing. In the soak method, one of the parts to be joined is soaked in ethylene dichloride for about 10 minutes or until a cushion has formed. This cushion should be deep enough to take care of any discrepancies in the fit of the parts. After the cushion has formed, press the wet surface against the opposite dry surface, and allow it to set for about a half a minute. Then apply enough pressure to the joint to squeeze out any air bubbles and assure complete intermingling of the cushions.

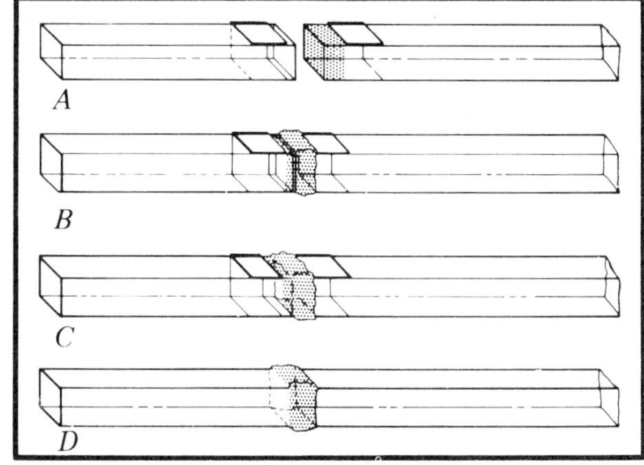

Fig. 6.5   The soak method of cementing acrylic plastics.

A.   *Mask off the plastic that is not to be affected by the solvent.*

B.   *Soak the edge of one piece in ethylene dichloride until a soft cushion has formed.*

C.   *Apply pressure to the two pieces so the already formed cushion will diffuse into the other piece, forming a cushion there.*

D.   *Allow the pressure to remain until the solvent has evaporated from the cushions and they become hard. The excess material can be removed, and the repair dressed to conform to the original material.*

The glue method is used when it is not convenient to soak the edges of one of the pieces. In this procedure, shavings of acrylic plastic are dissolved in ethylene dichloride, making a thick syrup, or glue. This is spread on one of the surfaces and the parts are assembled. After allowing the glue to set for a few seconds, apply pressure to mingle the cushions.

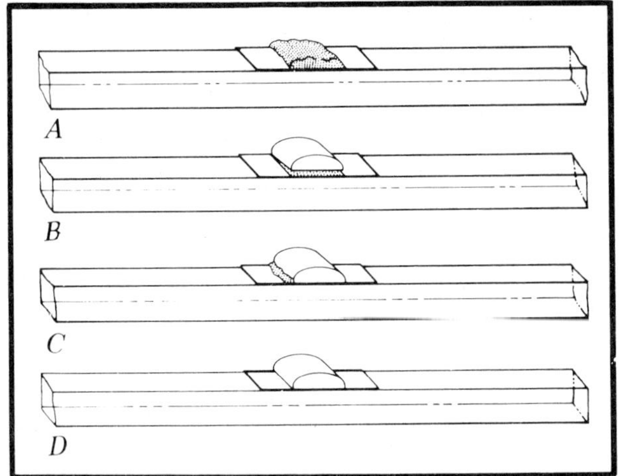

Fig. 6.6   *The glue method of cementing acrylic plastic.*

A.   *The areas not to be affected by the cementing process are masked off with pressure-sensitive aluminum tape, and a syrup made up of acrylic shavings dissolved in ethylene dichloride is spread on one of the surfaces.*

B.   *The piece to be cemented into place is laid into the syrupy glue, and pressure is applied.*

C.   *The excess syrup that oozes from between the two pieces of material is wiped off.*

D.   *The glue forms a cushion on both pieces, these cushions intermingle and a solid joint is formed.*

When applying any type of cemented patch on an acrylic plastic, make certain that the surface *not* to be affected is adequately protected from the adhesive. Pressure-sensitive aluminum tape should be used to mask the area around the repair.

b.   Application of pressure

It is of extreme importance that, once the parts to be joined have been assembled and the cushions adequately formed, an even pressure be applied to the joint. When the acrylic absorbs the adhesive, forming the cushions, it expands; however, when the adhesive evaporates from the intermingled cushions, the acrylic shrinks. The pressure that must be applied to assure that the cushions are properly intermingled must be great enough to force any air bubbles out, but not enough to break down the cushions or to cause localized stresses which would cause crazing.

This pressure must be applied by some method that, while allowing the cushion to shrink, will continue to *hold* the pressure. Spring clamps, weights, or rubber pads, are acceptable for this; parallel clamps or C-clamps, however, will loosen as the joint shrinks.

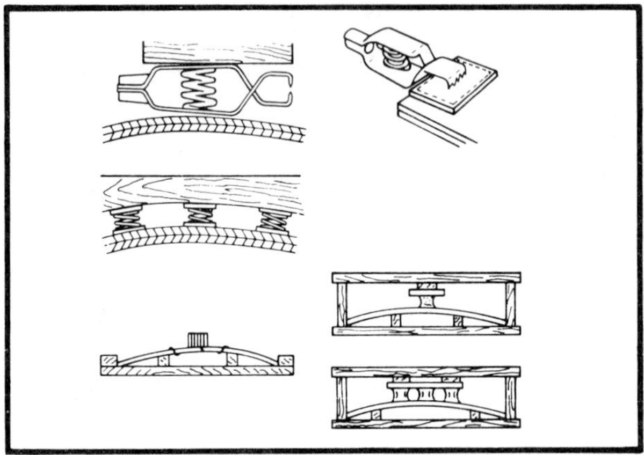

Fig. 6.7   *Pressure for curing cemented acrylics may be applied by a spring clip such as a battery clip, by coil springs, by sponge rubber, or by weights applied to the repair. C-clamps or parallel clamps are not satisfactory as they will loosen when the cushions shrink as the solvent evaporates.*

QUESTIONS

45.   What solvent is used as a cement for acrylic plastic?

46.   What is the syrup comprised of which is used for the glue method of cementing acrylic plastics?

47.   What can be used to mask an area of acrylic plastic that must *not* be affected by the solvent?

48.   Why are parallel or C-clamps not satisfactory for applying pressure when cementing acrylic sheets?

c. Curing

The solvent in a cemented joint will never
completely evaporate from the acrylic or the
cushions as they form, and since the cushions are
expanded, they will be weaker than the original
material. If the cemented material is heated to
around 122°F. (50°C.) and held at this temperature for about 48 hours, the cushions will expand.
The solvent contained will then be diffused into a
larger volume of plastic, becoming less concen-
trated, and giving the material more strength.

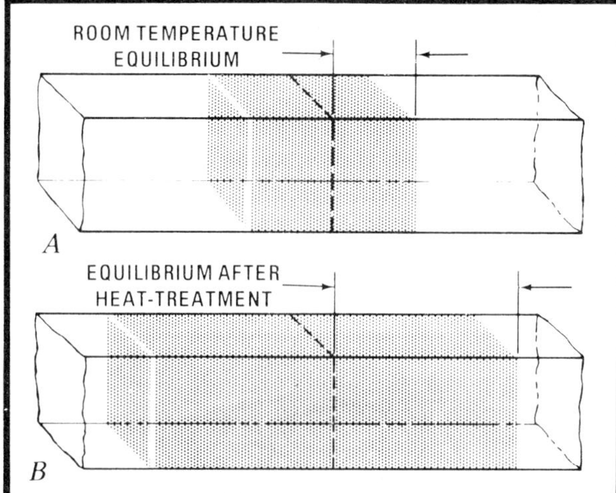

*Fig. 6.8   Heat treatment of an acrylic plastic
repair.*

> A. *The solvent never completely
> evaporates from the repaired area,
> and the plastic containing the
> solvent is weaker than that without
> the solvent.*
>
> B. *By heating the repaired area, the
> volume of entrapped solvent can
> be disbursed throughout a larger
> volume of plastic and the strength
> of the repaired area will be
> brought closer to that of the
> original material.*

Curing often called *heat treating*, will harden the
surface enough that machining or polishing can be
done without the danger of encountering soft
spots and resulting in an uneven surface.

QUESTION

49. What is meant by heat-treating a cemented
acrylic joint, and how does it increase the
strength of the joint?

D. Repairs

Aircraft windshields and side windows are
constructed of acrylic plastics. When damaged,
they are usually replaced unless the problem is
minor and does not require a repair that would be
in the line of vision. The reason for this is the high
cost involved in such repair and the fact that
replacement parts are readily available.

1. Temporary repair

There are times however when a windshield
may be cracked and must be put in good enough
condition to fly to where replacement can be
made. In a case such as this, temporary repair can
be made by stop-drilling the ends of the cracks
with about a number 30 drill, preventing the
concentration of stresses which would continue to
crack the windshield. A series of number 40 holes
a half-inch from the edge of the crack would then
be drilled, about a half-inch apart, and laced
through with brass safety wire.

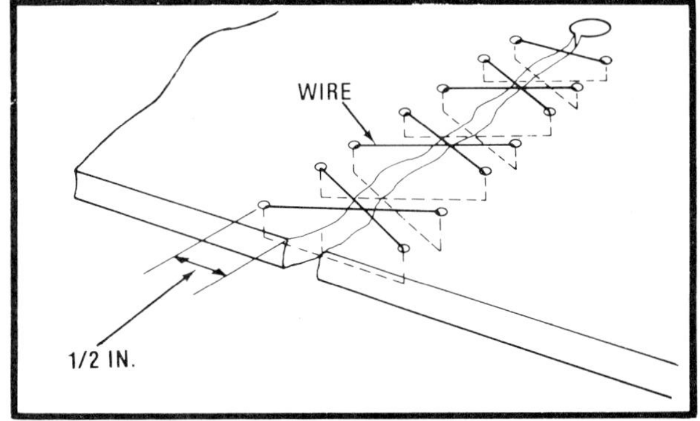

*Fig. 6.9   A temporary repair can be made to a
cracked windshield by stop-drilling the
ends of the crack, drilling holes on both
sides of the crack, and lacing the crack
together with brass safety wire.*

A second method of temporary repair is this:
Stop-drill the ends of the crack, and drill number
27 holes every inch of so in the crack.   Use
AN515-6 machine screws and AN365-632 nuts,
putting AN960-6 washers on both sides of the
plastic.   This will hold the crack together,
preventing further breakage until the windshield
can be properly repaired or replaced.

2. Permanent repair

Windshields or side windows with small cracks
that affect just the appearance rather than the
airworthiness of a sheet may be repaired rather

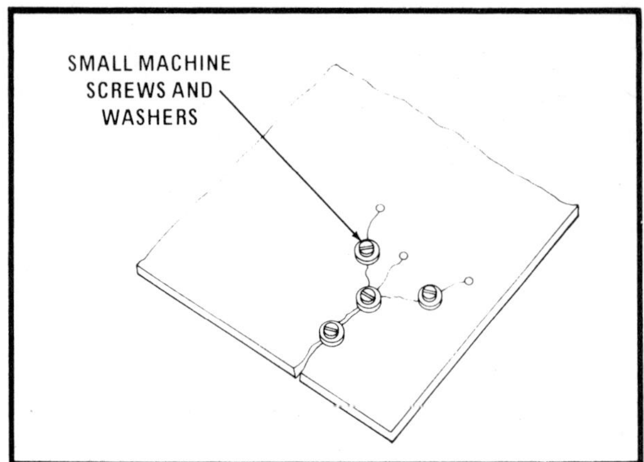

Fig. 6.10 *A temporary repair can be made to a cracked windshield by stop-drilling the ends of the cracks, drilling several holes along the crack, and using machine screws, washers and self-locking nuts to clamp the crack together.*

than replaced. Stop-drill the ends of the crack with a number 30 drill to relieve the stresses and spread them out.

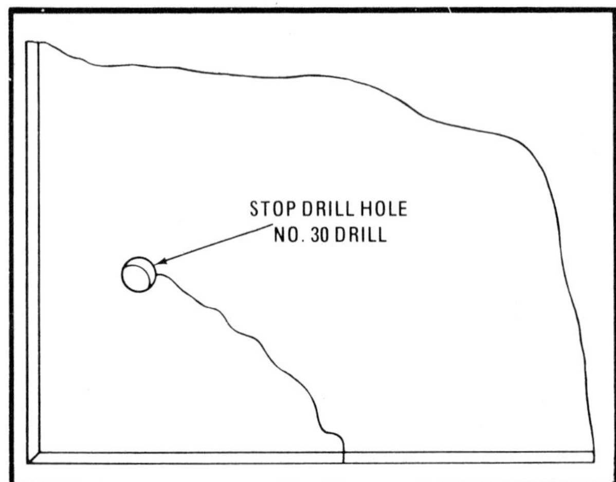

Fig. 6.11 *Permanent repairs may be made to acrylic windows by stop-drilling the ends of the crack, inserting the proper size acrylic dowel [which has been softened with ethylene dichloride], and flowing some of the solvent along the crack with a hypodermic needle.*

Using a hypodermic needle, fill the crack with ethylene dichloride, allowing capillary action to fill the crack completely. Soak the end of a 1/8'' (3.17mm) acrylic rod in ethylene dichloride to form a cushion and insert it in the stop-drilled

hole. Allow the repair to dry for about half an hour and trim the rod off flush with the sheet.

3. *Polishing and finishing*

Scratches and repair marks with certain limitations, can be removed from acrylic windows. No sanding should be done on any portion of a windshield that could adversely affect its optical properties, distorting the pilot's vision. Windows on pressurized aircraft must not have their thickness reduced to such extent by any repair that would weaken them. The manufacturer's Service Manual will specify the maximum amount they may be reduced in thickness.

If there are scratches or repair marks in an area that can be sanded, they may be removed by first sanding the area with circular rubbing motions. Use 320 to 400 abrasive paper wrapped around a felt or rubber pad.

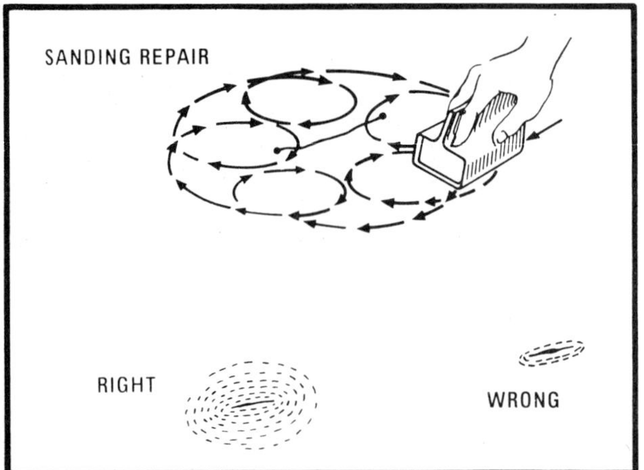

Fig. 6.12 *Scratches may be removed from acrylic sheets by sanding them in a circular pattern with fine sandpaper lubricated with mild liquid soap solution.*

Use light pressure and a mild liquid soap solution to serve as a lubricant. After this sanding is complete, rinse thoroughly with running water. Then, using a finer grit paper, continue to sand lightly, *always* rinsing until all of the sanding or repair marks have been removed.

After using the finest abrasive paper, apply a rubbing compound, buffing in a circular motion to remove all traces of the sanding.

4. Cleaning

Acrylic windshields and windows may be cleaned by washing with mild soap and much running water, rubbing the surface with your bare hands in the stream of water. After all of the dirt has been flushed away, dry the surface with a soft clean cloth or tissue and polish with one of the windshield cleaners especially approved for aircraft transparent plastics. (These may be purchased through aircraft supply houses.)

A thin coating of wax will fill any minute scratches that may be present. It will also cause any water coming in contact with the windshield to form droplets that are easily blown away by the wind.

E. Protection

Acrylic windshields are often called "lifetime" windshields, to distinguish them from those made of the shorter-lived acetate material. But even acrylics must be protected from the ravages of the elements.

If an airplane is parked in the direct rays of the sun, its windshield will absorb heat and will actually become hotter than either the inside of the airplane or the outside air. The sun will cause the inside of a closed airplane to become extremely hot, and with this heat also radiating onto the plastic windshield, the sun's damaging effect will be accelerated.

To protect against this, it is wise to keep the airplane in a hangar, or under the shade of a T-hangar. If this is not possible, some type of shade should be rigged to keep the sun from coming in direct contact with the windshield. Some aircraft owners have used a close fitting opaque, reflective cover over the windshield, but in many instances, this has done more harm than good. This cover will absorb moisture from the air and give off vapors. If it touches the surface of the plastic, it could cause crazing or minute cracks in the windshield. Another hazard in using such a cover is that sand could be blown under the cover, and the plastic could be scratched.

F. Installation

Aircraft windshields may be purchased either from the original aircraft manufacturer or any of several FAA-PMA (Federal Aviation Administration-Parts Manufacturing Approval) sources. These windshields are formed to the exact shape required, usually slightly larger than needed to allow them to be trimmed to exact size.

After the damaged windshield has been removed and all of the sealer cleaned from the grooves, the new windshield is cut to fit. This windshield, when first received, is covered with either a protective paper or film which prevents damage from handling or installation. Carefully peel back just enough of this covering to make the installation. The windshield should fit in its channels with an eighth of an inch clearance to allow for expansion or contraction. If there are any holes drilled in the plastic for screws, they should be about an eighth of an inch oversize.

Place the proper type of sealing tape around the edges of the windshield and install it in the frame. Screws that go through the windshield should be tightened down snug and then backed out a full turn, allowing the plastic to shift as it expands and contracts.

Do not remove the protective paper or film until the windshield is installed and the securing screws or rivets are in place.

QUESTIONS

50. What is one of the best solvents to use when cleaning an acrylic windshield?

51. What may be used to cover tiny scratches in a windshield, and to prevent rain sheeting-out and distorting the pilot's vision?

52. What are some of the disadvantages of using a close-fitting cover over the windshield of an airplane parked out in the rain?

53. Should a replacement windshield fit tightly into the channel in which it is installed?

Glossary

abrasive   A material containing minute particles of some substance which will tend to wear any surface which they rub.

accelerator   A substance added to a catalyzed resin to speed up the time required for the resin to cure.

acetone   A flammable liquid ketone (C_3H_6O) used as a solvent and a constituent in many aircraft finishes.

acrylic   A glossy, transparent thermoplastic used for cast or molded parts such as aircraft windshields.

adhesive   A substance used to bond two materials together by chemical means.

agent, bonding   An adhesive used to bond structural parts together.

agent, parting   A material or substance used to cover a mold and prevent the resin adhering to it.

agent, thixotropic   A substance added to a resin to increase its resistance to flow.

bond, chemical   The joining of two or more parts or pieces by molecular attraction of an adhesive agent which wets the parts.

brazing   The joining of metals by wetting their surface with a molten, non-ferrous filler metal. Brazing alloys usually have a melting temperature of above 800°F.

bond, mechanical   The joining of two or more parts or pieces by mechanical methods such as bolts, rivets or pins.

bonded structure   Structure joined together by chemical methods rather than mechanical.

catalyst   A material added to a resin to permit it to cure.

Cherry rivets   A form of blind rivet manufactured by the Cherry Rivet Division of Townsend. Its upset head is formed by pulling its tapered stem through its hollow shank.

corrosion   A chemical action caused by dissimilar metals acting in the presence of an electrolyte.

cure time   The time required for a resin to complete its solidification.

encapsulate   To completely enclose a component, or the fibers of a cloth, in a resin.

epoxy   A flexible, thermosetting resin with very little shrinkage upon cure. Its chief advantage is its ability to bond to a wide variety of surfaces.

fiberglass   Extremely thin fibers of glass. They may be woven into a cloth or lightly packed into a mat and are used to reinforce epoxy or polyester resins for aircraft structures.

filler   A material added to a resin to increase its bulk.

glue   An adhesive, primarily a gelatinous substance, that is capable of sticking materials together as it dries.

honeycomb   A hexagonal cellular material made of thin metal, paper, or plastic. It is used as a core material for sandwich structure.

inhibited sealer   A material used to exclude moisture and air from a honeycomb repair. In addition to sealing, it inhibits the formation of corrosion.

inhibitive film   A film of material on the surface of a metal which inhibits or retards the formation of corrosion. It does this by providing an ionized surface which will not allow the formation of corrosive salts on the metal.

inhibitor   An agent added to a resin to retard its curing and increase its shelf life.

life, pot The usable life of a resin. The time before it begins to thicken, after the catalyst and accelerator have been added.

life, shelf The useful life of an uncatalyzed resin.

MEK Methyl ethyl ketone. An important, low-cost solvent, similar to acetone. It is used as a cleaning agent to prepare a surface for painting, and as a stripper for certain finishes.

micro-balloons Miscroscopic size phenolic or glass spheres used to add body with very little weight to a resin when used as a filler or potting compound.

monomer A chemical compound that can be polymerized.

Plexiglas A proprietary name for a transparent thermoplastic acrylic material used for aircraft windows and windshields.

plywood Layers of wood glued together so that the grain in each layer is placed 45° or 90° to the other.

polyethylene A lightweight thermoplastic material that has very good chemical and moisture resistant characteristics. It is used for plastic sheeting and containers.

polymerization A chemical action in which two or more small molecules combine to form large molecules with the same structure as the original molecules.

potting compound A resin with filler capability used to fill honeycomb cells when making minor repairs to damaged honeycomb panels.

radome A plastic enclosure for the radar antenna.

resin A family of natural or synthetic fluids or semi-solid materials which may, by the addition of appropriate catalyzers, be changed into a solid.

resin, polyester A synthetic resin, usually reinforced with fiberglass cloth or mat, and used to form complex shapes of aircraft structures.

resin, thermoplastic A resin material which will soften with the application of heat. Most windshields and side windows are made of this material.

resin, thermosetting Most widely used in today's plastics, it sets usually by chemical means and maintains its hardness even when heat is applied.

sandwich construction A form of bonded structure in which a core material—such as metallic or plastic honeycomb, or end-grain balsa wood—is bonded between two sheets of metal or fiberglass sheet.

screeding tool A tool used to smooth out or level resins used in bonded structure manufacture or repair.

sealant A material used to form a seal between two imperfectly fitting surfaces. Sealants differ from gaskets in that they are usually liquid or semi-solid.

styrene A liquid hydrocarbon used in the manufacture of certain synthetic resins to improve their workability.

toluol The commercial grade of toluene which is a liquid aromatic hydrocarbon similar to benzene, but less volatile, flammable, or toxic.

welding A method of joining materials in which a portion of each piece is melted. The portions are then combined while in their molten state. Filler material is usually added for extra mass at the joint.

wood, balsa The light, strong wood of a tropical tree. When it is sliced across its grain and sandwiched between two face sheets of thin metal or fiberglass, it forms rigid, lightweight panels.

wood, laminated Layers of wood glued together so that the grain in each layer runs in the same direction.

woven roving A slightly twisted roll of glass fibers used to reinforce resins in making molded parts.

Answers To Study Questions

1. Resins will not adhere to, nor encapsulate glass fibers which have any oil on their surface.

2. The coarse-weave fibers allow the resins to encapsulate them better than the fibers in a fine-weave cloth.

3. Fiberglass mat.

4. A loosely twisted strand of glass fibers.

5. Paint cures by the drying, or evaporation, of the oils and solvents. Polyester cures by a chemical reaction.

6. Styrene thins polyester resins and makes them more manageable.

7. Catalysts suppress the action of the inhibitors and allow polyester resins to cure into a solid mass.

8. Accelerators speed the curing action of the resin and the catalyst.

9. A very heavy layer.

10. It shrinks.

11. 1. Follow the manufacturer's instructions in detail.

    2. Use a complete system made by one manufacturer. Do *not* mix brands.

12. Not below 65° nor above 85°F.

13. 1. It has a low percentage of shrinkage.

    2. It has high strength for its weight.

    3. It is exceptionally resistant to chemicals.

    4. It will adhere to an almost limitless variety of materials.

14. Thixotropic agents provide body to a resin, preventing its running as temperature thins it.

15. To remove all of the air from the resin, and to allow the atmospheric pressure to force the laminations together.

16. The critical side of the finished product. If the outside is critical, use a female mold, if the inside is critical, use a male mold.

17. Parting agents prevent the resins from sticking to the mold.

18. Honeycomb panels have much more rigidity.

19. Furnace brazing.

20. Balsa is sliced end-grain and sandwiched between sheets of aluminum alloy or fiberglass.

21. Solidly bonded structure produces a ringing sound when tapped, but a delamination will change this into a dull thudding sound.

22. Abrupt cross-sectional changes produce stress risers that concentrate the stresses and induce failure.

23. Seal all of the edges of the repair with a corrosion inhibiting sealant, and seal all air away from the interior of the panel.

24. Those made of Teflon or polyvinylchloride.

25. Methyl Ethyl Ketone (MEK).

26. Cellophane or polyvinyl alcohol film.

27. Potting compound. This may be mixed with micro-balloons to add body with minimum weight.

28. An insert plug of either balsa wood, or honeycomb material of the same density as that removed.

29. Only the type paint specified by the aircraft manufacturer.

30. Shearing is more efficient and cleaner than sawing, and it does not produce the dust which contains tiny particles of glass.

31. Acetone or Methyl Ethyl Ketone (MEK).

32. Be sure that the lamps are not placed so close to the surface that they overheat the repair and damage the polyethylene sheet.

33. Be sure that the entire repair is sealed so moisture cannot enter it and cause corrosion.

34. A doubler can be bonded or riveted over the dented area.

35. Acrylic.

36. Acetone rubbed onto acrylic will turn white, but it will soften acetate plastic.

37. Support them in a tilted position, about ten degrees off of vertical. In this way they will not buckle, and danger of scratching will be minimized.

38. It will craze. This means that thousands of tiny cracks will appear on the surface that is subjected to the tensile stress.

39. It will cause the acrylic to turn milky or cloudy.

40. 110°C. or 230°F.

41. The teeth should be side-dressed after they have been set.

42. The material is being fed into the saw too fast, or the blade is improper for the material.

43. Water-soluble cutting oil.

44. It should have a greater included angle.

45. Ethylene dichloride.

46. Acrylic shavings dissolved in ethylene dichloride.

47. Pressure-sensitive aluminum tape.

48. Clamps of these types will loosen as the cushions shrink from the solvent evaporating.

49. Heat-treating is done by raising the temperature of the repair so the entrapped solvent can diffuse into a larger volume of the material. The lower the concentration of entrapped solvent, the stronger the material.

50. Lots of clean running water.

51. A thin coating of wax.

52. Close-fitting windshield covers entrap heat, absorb moisture, and release fumes that cause crazing of the windshield.

53. No. At least one-eighth of an inch should be allowed for expansion of the windshield.

Bonded Structure

Final Examination

STUDENT _______________________________

GRADE _______________________________

Place a circle around the letter for the answer which is most nearly correct.

1. What is the purpose of adding the catalyst to a polyester resin?

    A. It increases the shelf life of the resin.
    B. It aids in joining together the polyester molecules.
    C. It prevents the resin hardening.
    D. It makes the cured resin more flexible.

2. What actually causes a polyester resin to cure?

    A. Evaporation of the solvents
    B. Oxidation of the accelerator
    C. Solidifying of the inhibitors
    D. Heat

3. Which of two identical batches of polyester resin will cure faster?

    A. One spread in a thin layer
    B. One spread in a thick layer
    C. One kept in an airtight container
    D. One kept in a vented container

4. Which statement is *not* true about epoxy resin?

    A. Epoxy will not stick to glass.
    B. Epoxy resins may be colored.
    C. Some epoxy resins are quite thin; others are thick.
    D. Epoxy resins have a low precentage of shrinkage.

5. What may be used to clean brushes or tools that have uncured epoxy resin in them?

    A. Acetone
    B. Methyl ethyl ketone
    C. Unleaded gasoline
    D. Di-acetone alcohol

6. What is the function of micro-balloons added to epoxy resin?

    A. They add volume but little weight to the resin.
    B. They allow the resin to cure at a lower temperature.
    C. They increase the weight of a given volume of resin.
    D. They furnish the required color to the resin.

7. What is gained when a honeycomb material is used in place of sheet metal in certain structural applications?

    A. Tensile strength
    B. Resistance to abrasion
    C. Rigidity
    D. Lower construction costs

8. What is a good method of determining the extent of delamination in a metal-faced honeycomb structure?

    A. Shine a light through the panel and look for a change in density.
    B. Tap a coin along the damage, and listen for a change in sound.
    C. Look up this type of damage in the Type Certificate Data Sheets.
    D. There is no way to tell; the entire panel must be replaced.

9. When making a repair to a honeycomb structure, what should be done to the damaged core?

    A. Remove it and leave it out.
    B. Remove the damaged core material and put in a plug of heavier honeycomb. The plug need not be bonded in place.
    C. Leave the damaged honeycomb where it is, but put on new face sheets.
    D. Remove the damaged core material, and bond in a plug of the same density honeycomb as was removed.

10. What consideration must be taken when repairing a radome that is not required of other honeycomb structures?

    A. Do not make abrupt changes in thickness.
    B. Do not make any repair that will change its electrical properties.
    C. Do not make any repairs that will cause abrupt changes in the airflow.
    D. Do not bond the face sheets to the core.

11. What precautions should be taken to prevent corrosion forming inside a repaired metal honeycomb structure?

    A. Prime the repair with a corrosion-inhibiting primer and seal the atmosphere from the repaired area.
    B. Paint the repaired structure to match the color of the rest of the airplane.
    C. Be sure the inside of the repair is vented to the outside air to prevent moisture condensing.
    D. Do not use heat to cure the repair, as it will cause moisture to be drawn into the repaired area.

12. What is the best way to be sure that the resin for a bonded repair is properly mixed?

    A. Mix in proportions according to weight, not volume.
    B. Mix in proportions according to volume, not weight.
    C. Mix to a consistency shown to be correct by the use of a viscosimeter.
    D. Make a test sample using a bit of the resin you have mixed.

13. If an airplane must be parked in the hot sun, what is the best way to protect the windshield?

    A. Cover the windshield with a close-fitting opaque windshield cover.
    B. Don't worry about it, direct sunlight cannot damage acrylic plastics.
    C. Lay a close-fitting paper or cardboard cover over the windshield.
    D. Rig some form of shade that will protect the windshield from the direct rays of the sun.

14. What type of repair can be made to a dented honeycomb face skin if the skin is required to have its original strength?

    A. Fill the dent with Bondo, sand it smooth, and finish it to match the rest of the structure.
    B. Remove the area around the dent and rivet a patch over the hole.
    C. Bond a doubler of the same material and at least the same thickness as the dented skin over the dent.
    D. Do nothing to repair it, as dents do not decrease the strength of the face skin.

15. If a balsa wood plug is used to replace some of the damaged honeycomb, how should the plug be cut?

    A. So the fibers of the wood are parallel with the face skin.
    B. So the fibers of the wood are perpendicular to the face skin.
    C. The balsa is shredded and mixed with resin to fill the hole.
    D. Balsa cannot be used to replace honeycomb material in a repair.

16. What type of repair is recommended for a honeycomb structure when it is felt that sufficient control cannot be exercised to *assure* an adequate bond?

    A. Replace the entire bonded structure.
    B. Go ahead and make the bonded repair, but be extremely careful.
    C. A riveted repair, using structural blind rivets.
    D. Adequate control is easy to maintain, so bonded repairs are always assured of having adequate strength.

17. When making a riveted repair to a honeycomb structure, what step should be taken to prevent corrosion starting around the rivets?

    A. Use only corrosion-resistant steel rivets.
    B. Use only pure aluminum rivets.
    C. Dip the rivets in corrosion-inhibiting sealer and install them while they are still wet.
    D. Install the rivets in a conventional manner, and paint the repair with multi-coats of paint.

18. What type of repair should be made to a damaged bonded structure if there are no specific details in the manufacturer's service manual?

A. Only those repairs described specifically in the service manual can be made to a bonded structure.
B. Only repairs detailed in Advisory Circular 43.13-1A can be used to repair a bonded structure.
C. If the repair is not described, you can devise your own.
D. A repair that has been approved by the manufacturer's engineering and service departments.

19. What is recommended as a cement for repairing acrylic plastic?

A. Methyl Ethyl Ketone
B. Ethylele dichloride
C. Acetone
D. Unleaded gasoline

20. Which process is recommended for cleaning an aircraft acrylic windshield?

A. Wipe it carefully with a dry shop towel.
B. Clean it with aircraft rubbing compound.
C. Wash it with mild soap and running water.
D. Wash it with clean paper towels saturated with unleaded gasoline.

Bonded Structure

Answers to Final Examination

1.	B	11.	A
2.	D	12.	D
3.	B	13.	D
4.	A	14.	C
5.	D	15.	B
6.	A	16.	C
7.	C	17.	C
8.	B	18.	D
9.	D	19.	B
10.	B	20.	C

Appendix

APPENDIX A

*MANUFACTURERS OF BONDED STRUCTURE
AND BONDED STRUCTURE MATERIAL*

Aerospace Division of UOP
Route 202
Bantam, Connecticut 06750

Airline Systems
Division of Adhesive Engineering Company
1411 Industrial Road
San Carlos, California 94070

Ceconite, Inc.
4677 Worth Street
Los Angeles, California 90063

Essex Chemical Corporation
Coast Pro-Seal Division
19451 Susana Road
Compton, California 90221

Furane Plastics Incorporated
5121 San Fernando Road
West Los Angeles, California 90039

Gates Engineering Company, Inc.
100 S. West Street
Wilmington, Delaware 19899

Gem-O-Lite Plastics Corporation
5525 Cahuenga Boulevard
North Hollywood, California 91601

M.C. Gill Corporation
4056 Easy Street
El Monte, California 91731

Occidental Petroleum Corporation
Hooker Durez Division
North Tonawanda, New York 14120

Pennwalt Lucidol
1740 Military Road
Buffalo, New York 14240

Products Research and Chemical Corporation
2919 Empire Avenue
Burbank, California 91504

REN Plastics, Inc.
Lansing, Michigan 48909

One of the nation's most complete repair
facilities for bonded structure is:

Nordam
A Division of R.H. Siegfried, Inc.
510 South Lansing
Tulsa, Oklahoma 74120

APPENDIX B

The following adhesives **and sealers do not**
constitute a comprehensive list; **they do, however,**
provide basic information **on some of the available**
bonded structural materials.

*Due to the wide variety of uncontrolled conditions,
these are only suggested adhesives. **We urge**
testing to determine the suitability of adhesives
for specific applications.*

[A] *Adhesive Engineering*
    1411 Industrial
    San Carlos, California 94070

[B] *Gates Engineering*
    100 S. West Street
    Wilmington, Delaware 19899

[C] *Furane Plastics Inc.*
    5121 San Fernando Road
    West Los Angeles, California 90039

[D] *Essex Chemical Corporation*
    19451 Susana Road
    Compton, California 90221

[A] *AS-401-1*
**Two-component rapid-curing epoxy adhesive
system, specially formulated for quick-fix field
repair of damaged honeycomb panels and metal
or fiberglass facings.**

[A] *AS-410*
Two-component resin system, primarily used for wet lay-up repair of laminated plastic structures. This system may also be used for certain metal repairs.

[C] *EPIBOND 122*
A gray, viscous paste, combined with a curing agent, serves as a general purpose epoxy adhesive, well suited for bonding metals, plastics, wood, and related materials.

[C] *EPIBOND 1210*
A general purpose adhesive that may be combined with various hardeners to bond metals, plastics, wood, and related products. Because of its semi-fluid consistency, it is well suited to application by brush, spatula, or dispensing units. It readily bonds to many diverse materials.

[C] *EPIBOND 1217-A/B*
A clear amber epoxy paste adhesive, capable of a rapid cure at 75°F. It is recommended for bonding steel, aluminum, wood, ceramics, and a variety of plastic substrates.

[C] *EPIBOND 8510 A/B*
This is a metal bonding adhesive which meets the requirements of MM-A-132, Type I, Class 3, and MIL-A-5090-D, Type I. It possesses good bond strength from $-65°$ to $+300°$F. and is resistant to creep, rupture, and fatigue. It cures hard overnight at room temperature.

[C] *EPIBOND 87399-A/B*
A two-part, smooth epoxy resin paste with an amine-type curing agent. Meets MM-A-132, Type I, Class 3 specifications.

[B] *GACO N-83*
A thermally reflective, rain-erosion-resistant compound meeting MIL-C-27315, Class 1 specifications.

[D] *PRO-SEAL 826*
Integral fuel tank sealant, fuselage sealant, pressure cabin sealant, oil-resistant sealant. Meets MIL-S-7502B and MIL-S-11031B specifications.

[D] *PRO-SEAL 444R*
Fuel tank top coating and sealant (Buna-N rubber based). Meets MIL-S-4383B specifications.

[D] *PRO-SEAL 509*
Low-adhesion gasketing sealant. Two-part, black, flexible, hand strippable gasket sealant. Meets MIL-C-15705A specifications.

[D] *PRO-SEAL 567*
Void-filling compound (polysulfide base). Permanently flexible, low-shrinkage void filler. Meets BMS 5-16A specifications.

[D] *PRO-SEAL 584*
High-strength neoprene-based adhesive. Flexible, high temperature-resistant. Water-resistant. Meets MIL-A-5092A, Type II specifications.

[D] *PRO-SEAL 590M*
General purpose Buna-N adhesive. High-temperature-resistant adhesive. Excellent adhesion without priming to most surfaces. Excellent fuel and oil resistance. Retains strong bond at elevated temperatures. Meets MIL-C-4003 specification.

[D] *PRO-SEAL 700*
Heat-resistant firewall sealant/coating to 400° F., good adhesion to metals, permanent flexibility, self-extinguishing. Meets MIL-S-38249, Type I specifications.

[D] *PRO-SEAL 706*
Access door (low-adhesion) sealant and gasketing compound (polysulfide base). Meets MIL-S-8784A specifications.

[D] *PRO-SEAL 707*
Integral fuel tank sealant, pressure cabin and fuselage sealant; oil-resistant sealant. Meets MIL-S-7502C, all types specification.

[D] *PRO-SEAL 711*
Equal part sealant; integral fuel tank, pressure cabin and fuselage. Exceptional resistance to dissimilar metal corrosion. Meets MIL-S-750B, Class B-2 specifications.

[D] *PRO-SEAL 714*
Suitable to 400°F. use, intermittent to 500°F. Permanently flexible. Meets BMS-5-18, Type II Lockheed C-40-776, Rockwell, Int. 234-722523-4 specifications.

[D]   *PRO-SEAL 727*
Electrical potting and encapsulating compound, polysulfide base. Room temperature cure, good adhesion, good electricals from −65°F. to +200°F., outstanding weather-, gas-, oil-, and water-resistance; excellent hydrolytic stability. Meets MIL-S-8516E specifications.

[D]   *PRO-SEAL 735*
Aerodynamic smoothing sealant. Aluminum colored, thixotropic, two-part, flexible sealant for aerodynamic smoothing. Meets DMS-1819A specifications.

[D]   *PRO-SEAL 755*
Two-component, elastomer-based coating. Recommended prime coat for bonding elastomeric resin systems to flexible synthetic surfaces such as neoprene, PVC, nylon, treated Teflon, etc.

[D]   *PRO-SEAL 790*
Polyurethane electrical potting compound. Two-part polyurethane for electrical potting. Good electrical properties. Black only. Does not contain MOCA. Cure above 180°F. only. Meets ABO 130-113 specifications.

[D]   *PRO-SEAL 824*
Epoxy structural potting compound. One-part (frozen) thixotropic epoxy for pre-potting of honeycomb aircraft construction. High-strength properties to 400°F.

[D]   *PRO-SEAL 826*
Epoxy adhesive. One-part (frozen) liquid epoxy adhesive for elevated temperature use. Meets RMS-003, Type VII specifications.

[D]   *PRO-SEAL 828*
Epoxy structural potting compound—low density. One-part (frozen) thixotropic epoxy for pre-potting of honeycomb aircraft construction. High-strength properties to 350°F.

[D]   *PRO-SEAL 829*
Epoxy structural potting compound. One-part (frozen) thixotropic epoxy for pre-potting of honeycomb aircraft construction. High-strength properties to 400°F.

[D]   *PRO-SEAL 890*
Integral fuel tank sealant, pressure cabin sealant; jet fuel-resistant. Meets MIL-S-8802D specifications.

[D]   *PRO-SEAL 895*
Aluminum pigmented, aerodynamic smoothing compound (polysulfide base). Meets MIL-S-38228, Type I, and MIL-F-81108, Type I specifications.

[D]   *PRO-SEAL 896*
Void-filling compound (polysulfide base). Permanently flexible, low-shrinkage void filler; excellent heat-, fuel-, oil-, and water-resistance.

[D]   *PRO-SEAL 898*
Fuel-resistant integral fuel tank sealant, cabin pressure sealant, airfoil smoothing compound. Special light gray color. Exceptional strength and jet fuel-resistance.